PROGRESS IN COLLOID & POLYMER SCIENCE
Editors: H.-G. Kilian (Ulm) and G. Lagaly (Kiel)

Volume 87 (1992)

Solidification Processes in Polymers

Guest Editors:
J.-F. Jansson and U. W. Gedde (Stockholm)

Springer-Verlag
Berlin Heidelberg GmbH

ISBN 978-3-662-15690-2   ISBN 978-3-7985-1682-3 (eBook)
DOI 10.1007/978-3-7985-1682-3
ISSN 0340-255 X

Originally published by Dr. Dietrich Steinkopff Verlag GmbH & Co. KG, Darmstadt in 1992
Softcover reprint of the hardcover 1st edition 1992

Chemistry editor: Dr. Maria Magdalene Nabbe; English editor: James Willis; Production: Holger Frey.

Type-Setting: Graphische Texterfassung, Hans Vilhard, D-6126 Brombachtal

Preface

The proceedings of the 23rd Europhysics Conference on Macromolecular Physics is devoted to solidification of polymers in general. The proceedings constitute a collection of papers dealing with crystalline, liquid crystalline, and amorphous polymers, and also include oriented polymers and the effects of pressure and processing. The conference was held in Stockholm, May 27—31, 1991, and it was sponsored by Neste OY, Finland, and the National Swedish Board for Technical Development (STU). The conference was organized by the Royal Institute of Technology (Department of Polymer Technology), Stockholm, Sweden, and by the European Physical Society, Section of Macromolecular Physics. Members of the International Advisory Board were F. J. Baltá Calleja, Spain; A. Keller, United Kingdom; D. C. Bassett, United Kingdom; A. J. Kovacs, France; E. W. Fischer, Germany, and L. C. E. Struik, Netherlands. The local organizing committee consisted of U. W. Gedde (chairman), B. Asker, A. Gustafsson, B. Stenberg, B. Terselius, A.-C. Albertsson, J. Kubat, J.-A. Månson, and B. Törnell. Prof. J.-F. Jansson was Chairman of the National Committee. The organization of the conference was also made possible by the efforts of the staff of the Department of Polymer Technology, Royal Institute of Technology.

We would like to express our appreciation to the authors of the papers, to Steinkopff Verlag, Darmstadt, Germany, and to the referees who read the papers with great care and improved the overall quality.

Stockholm, March 1992

Ulf W. Gedde
Jan-Fredrik Jansson

Contents

Crystalline polymers

Isotactic polypropylene: Growth transitions and crystal polymorphism

B. Lotz and J. C. Wittmann

Institut Charles Sadron (CRM-EAHP), CNRS-ULP Strasbourg, France

Abstract: The following recent results on growth transitions and crystal morphology of isotactic polypropylene (iPP) are presented: a) observation of a hitherto unsuspected low-temperature growth transition between β and α phases; b) structural characterization and experimental check of the γ phase which is made of chains with non-parallel axes, and c) observation by atomic force microscopy of the crystallographic (010) contact face epitaxially interacting with benzoic acid substrate.

Key words: Isotactic polypropylene; crystal phases; growth transitions; atomic force microscopy; epitaxy

Introduction

The crystallization behavior and crystal polymorphism of isotactic polypropylene (iPP) are among the most complicated in the crystalline polymer field. This complexity arises from the existence of four well-characterized crystal phases (α, β, γ, and smectic), out of which until recently [1], only the α phase had been analyzed in molecular detail. The structure of the γ phase has now been solved, and it is unique in that it involves nonparallel polymer chains [2]. The structure of the hexagonal β phase and, to some extent, of the smectic phase obtained on quenching have not yet been or are only partly solved [1]. The complexity is also linked to the specificity of crystallization conditions required to induce or promote formation of the four phases: whereas the α phase is formed spontaneously, the β phase is promoted by specific nucleating agents, the γ phase by the use of low molecular weight materials or crystallization under high pressure, and the smectic phase by fast crystallization on quenching.

The present contribution deals with three different investigations in the crystallization behavior or crystal phases of iPP in which our laboratory has been involved. They include: a) observation of hithertho unsuspected low-temperature growth transitions in iPP [3]; b) experimental check of the original non-parallelism of chains in the γ phase [4]; c) observation of the crystallographic face of α iPP interacting epitaxially with a low molecular weight substrate crystal [5]. Most of these works were reported recently, either in a final [4] or preliminary form [3, 5].

Experimental

Growth transitions have been observed by optical microscopy on thin film of iPP deposited on a glass cover slide and sprayed with γ chinacridone, a well known nucleating agent of the β phase of iPP. After complete melting, the film is allowed to crystallize at $\simeq 130\,°C$ for a few seconds and then deposited on a substrate at room temperature. Video recording of the growing spherulites observed in a polarizing microscope enables estimation of growth rates in the vicinity of the transition temperatures.

Single crystals of the γ phase are produced by thin-film growth of a thermally decomposed iPP, i.e., that which contains significant amounts of low molecular weight material ($\simeq$1000—2000). After shadowing, the crystals are examined using the rotation-tilting stage of a Philips CM12 electron microscope.

Epitaxially crystallized iPP [6] is produced by co-melting benzoic acid and iPP between two cover slides (iPP is first submitted to a light degradation in order to bring its crystallization range close to that of benzoic acid). After dissolution of benzoic acid, the sample is ready for examination with the atomic force microscope (Nanoscope II, conducted at and in collaboration with the Freiburger Materialforschungszentrum, Freiburg im Breisgau, FRG).

Results

Low T_c growth transitions in isotactic polypropylene

This work was prompted by our own results on the self-seeding of α and β phases of iPP [7] and by the results of Varga and Toth [8] on interplay of thermal histories and crystal polymorphism of iPP. Specifically, Varga and Toth report that the crystal phase formed after partial melting of an initial β phase may be either α or β, depending on the crystallization *and post crystallization* history of the initial β phase. If the β phase is taken below $T^* \simeq$ 100 to 105 °C at any stage of the thermal cycle (even *after* isothermal crystallization) an α phase is produced on recrystallization; if T remains always $>T^*$, a β phase is produced. This behavior suggests development of numerous α phase nuclei at a temperature $T <$ 105 °C which, being thermally stable, saturate the crystallization process on recrystallization after partial melting.

Experimental evidence in favor of this mechanism can be sought for by *physically separating* crystallization processes occurring at different temperatures, i.e., by examination of concentric rings formed during quench. As shown in Fig. 1, the bright β phase spherulites experience a growth transition to the α phase, typically much less birefringent as a result of lamellar branching [11]. At lower temperatures (i.e., on outer portions of the spherulites) a more birefringent ring is formed which can be either a smectic phase, or an α phase with reduced impact of lamellar branching and, therefore, negative birefringence of the spherulite. Recent electron diffraction experiments on partially etched spherulites of this type suggest that this ring is of the α phase. Finally, at temperatures close to 80 °C, onset of homogeneous nucleation results in the formation of a weakly birefringent background, possibly formed of the smectic phase (domain marked c). The major β to α transition revealed by this study is assumed to be located at $\simeq$105 °C, based on the results reported by Varga and Toth. It corresponds indeed to temperatures near the maximum of the $\log G = f(T_c)$ curve derived by Icenogle for iPP [9]. Growth rates measured near the growth transition are $\simeq$60 μm/second, i.e., close to those predicted by theory. These observations provide a physically plausible explanation for the puzzling results reported by Varga and Toth; they also point to the first example (to our knowledge) of a polymer crystal phase kinetically favored in a window of crystallization temperatures. Indeed, as shown by Varga [10], a β- to α growth transition also occurs at 141 °C, on the high T_c side of the crystallization range of iPP. As a result, the β phase is kinetically favored only in the 105—141 °C T_c interval.

Non-parallel chains in the γ phase of iPP: an experimental test

Recently, a highly original structure for the γ phase of iPP has been proposed [2]. It differs from all known polymer structures in that it comprises

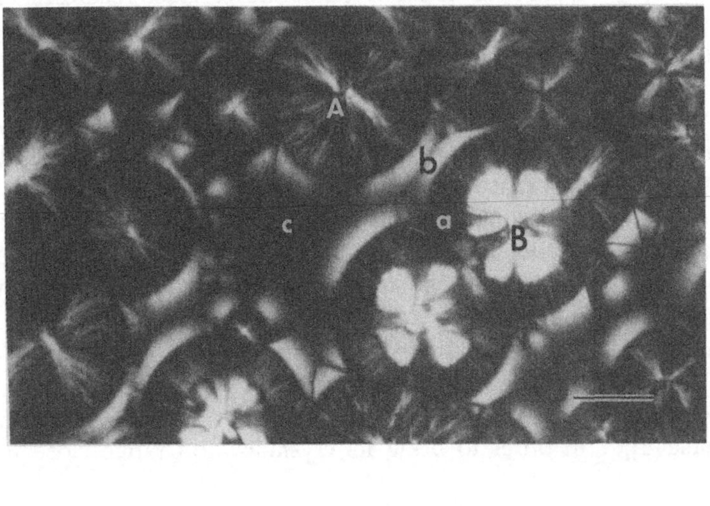

Fig. 1. Growth transitions revealed in iPP spherulites on quenching. Bright spherulite interiors (marked B) are of the β phase, and are nucleated by γ-chinacridone. Outer, darker rings are of α phase (marked a). A: α phase spherulites; b: negative birefringence α growth; c: homogeneous nucleation. Optical micrograph, crossed polars, scale bar: 20 μm. (Reprinted from [3].)

non-parallel chains: two sets of 3_1 helices are at an angle of 80°, i.e., are parallel to the diagonals of the *ab* face of an orthorhombic unit cell with parameters a = 8.54, b = 9.93, and c = 42.41 Å. This chain arrangement is a systematization of the branching mechanism of α phase crystals analyzed in molecular detail [11] and which gives rise to the well-known branching lamellae of quadrites [12]. The proposed γ phase structure rests, however, for the most part, on an analysis of its x-ray powder pattern by the Rietveld method [2].

Considering the originality of the structure, a more direct check of the existence of two-chain orientations has been devised. The experimental test rests on the exploration of the full reciprocal space of single crystals of γ iPP, and on comparison of diffraction patterns for zones axes parallel to two-chain axes. Single crystals of α and γ phase crystals were produced as in a previous study [13]. γ phase crystals branch on first-grown α phase crystals at a 40° angle (Fig. 2). However, they often topple over and lie flat on the film surface. These flat-on crystals provide the desired test material: + and —40° tilts

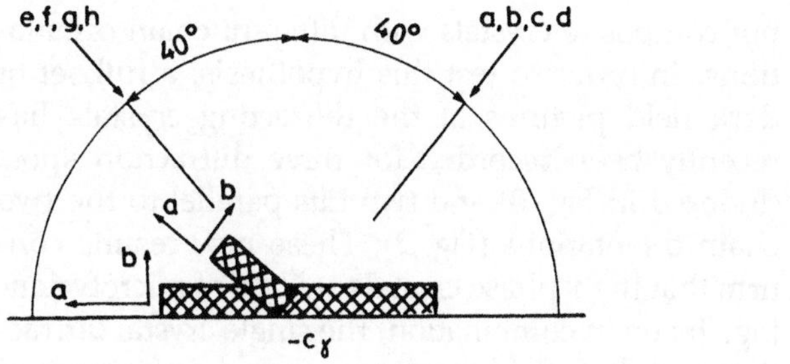

Fig. 2. Structure and molecular organization in γ phase single crystals grown in thin film as seen from a direction parallel to the film surface. Stripes correspond to chain axis orientations. Oblique crystals are nucleated on the elongated α phase single crystal "spine" in Fig. 3. Further growth may take place parallel to the substrate surface and yields the triangular crystals seen in Fig. 3. Viewing angles used in Fig. 3 are marked

in an electron microscope tilting stage produce similar diffraction patterns, as shown in Fig. 3 (left).

However, the above test is sufficient only if the single crystals are free of microtwinning, i.e., are

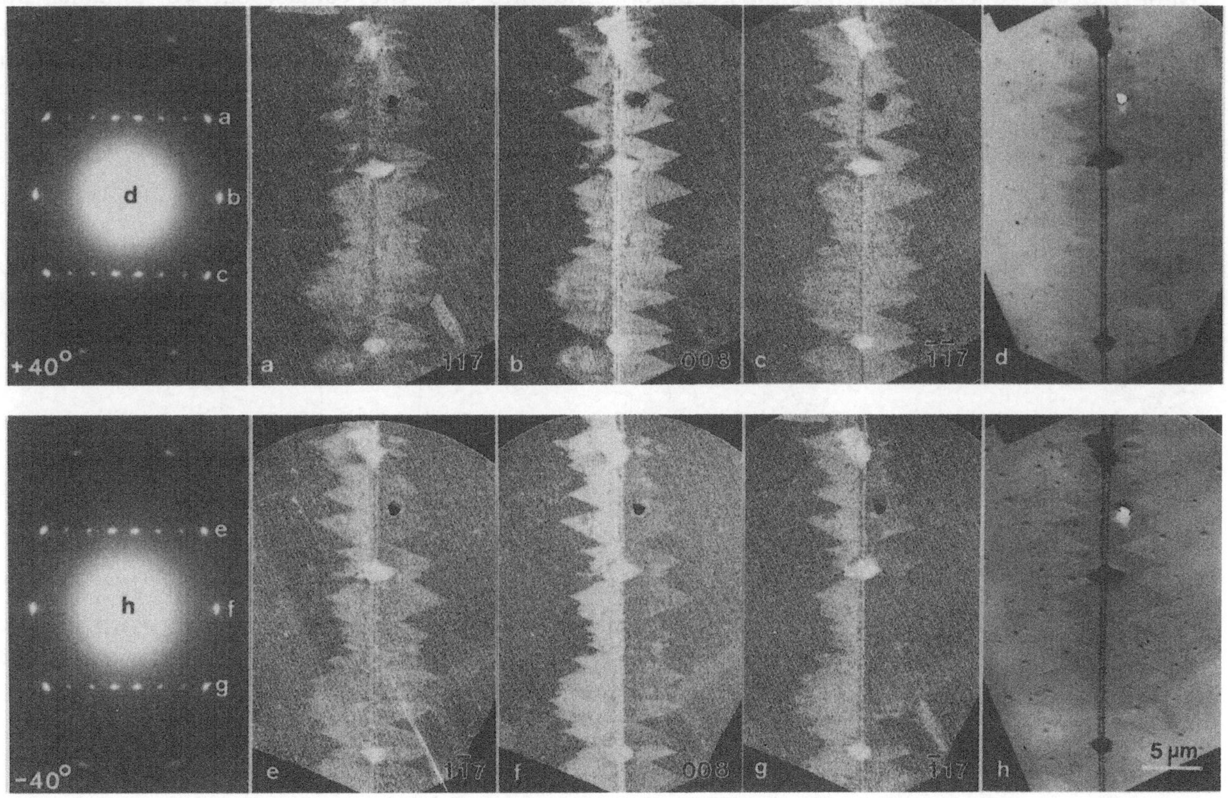

Fig. 3. Diffraction patterns, dark fields, and bright fields of composite α and γ phase iPP single crystal entities as observed from viewing directions indicated in Fig. 2. Dark and bright field pictures are from spots indicated in the diffraction patterns. Note the similarity of patterns taken with tilt angles at + and —40° (zone axes 110 and $\bar{1}$ 10)

not composite crystals with different chain orientations. In order to test this hypothesis, a full set of dark field pictures of the diffracting crystals has recently been recorded for three diffraction spots (indexed in Fig. 3) and two tilts parallel to the two chain orientations (Fig. 2). These new results confirm that the γ phase crystals are free of microtwinning. Taken in combination, the single crystal diffraction and dark field results confirm the existence of two chain orientations in γ iPP and, therefore, provide strong experimental support for the structure of Brückner and Meille [2].

AFM on epitaxially crystallized α isotactic polypropylene

Atomic force microscopy (AFM) provides a new and powerful tool for investigation with near atomic resolution of the surface structure of materials: for example, chain end methyl groups have been resolved in unfiltered images of the surface of paraffin crystals [14].

AFM is ideally suited to investigate the details of molecular and crystallographic interactions involved in epitaxial crystallization if and when the contact face can be exposed. In the case of polymer/low molecular weight organic substrates investigated in great detail in our laboratory, the crystallographic interactions are usually established by electron diffraction, often of composite polymer-substrate films [6]. This technique may, however, prove insufficient for fully determining these interactions: AFM can then be of help, as illustrated now for the epitaxial relationship between α iPP and benzoic acid [5].

Epitaxial crystallization of iPP on benzoic acid crystals results in an iPP contact plane (exposed after dissolution of benzoic acid) with a morphology very reminiscent of quadrites: two arrays of edge-on lamellae intersect at an angle of $\simeq 80°$ [6]: electron diffraction [11] indicates that the crystallographic plane in contact with the benzoic acid substrate is (010). However, from the crystal structure of α iPP, it is clear that two possible contact planes remain possible. These planes, marked A and B in Fig. 4a, differ by the density of exposed methyl groups, as shown also in Fig. 4b.

Atomic force microscopy makes it possible to discriminate between these two possibilities. Figure 5 shows a Fourier-filtered atomic force image which indicates that the contact surface contains the lowest density of methyl groups, i.e., is a (010) face of type B. Low resolution AFM pictures further reveal the lamellar surface orientation of the edge-

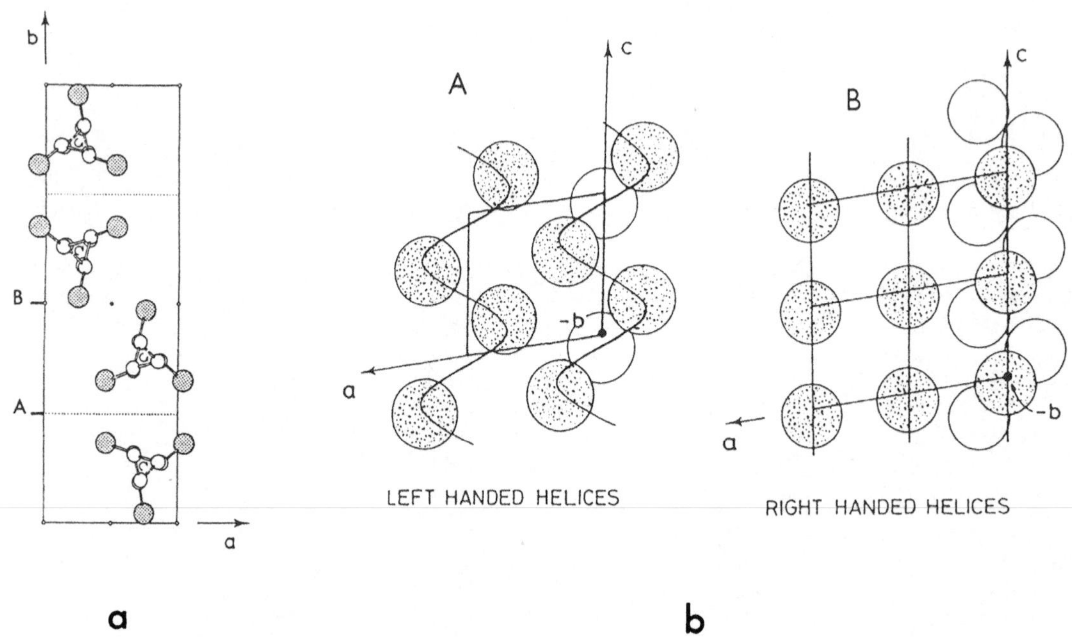

Fig. 4. a) Crystal structure of α phase iPP in c axis projection. Note the 3_1 helical conformation and existence of two different (010) planes, marked A and B. b) Disposition of methyl side chains in (010) planes A and B of α phase iPP; note the difference in methyl group density

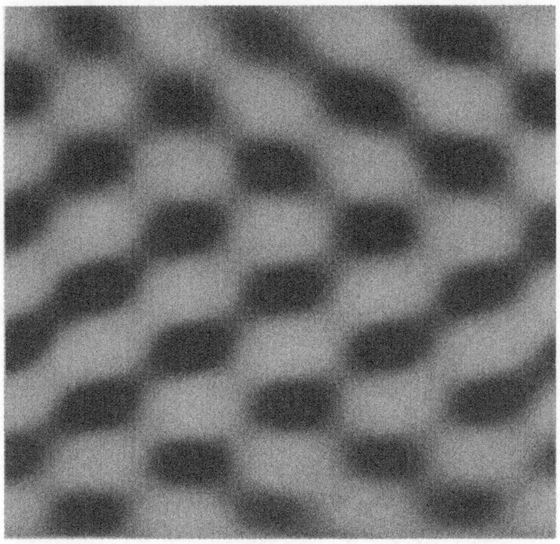

Fig. 5. Fourier-filtered force image of the (010) α phase iPP contact face with benzoic acid. Note the characteristic lozenge-shaped pattern of dark spots (methyl groups) indicating a (010) face of type B. Size of imaged area: 3 × 3 nm

on lamellae. Combining this information with the unit-cell orientation indicated by Fig. 5, it is possible (cf. [11] and, Fig. 9 therein) to determine the helix sense of the observed helices. In the area imaged in Fig. 5, they are righthanded.

Finally, very recent results relate to the epitaxy of γ iPP on benzoic acid. The γ phase is identified by its characteristic relative orientation of chain axes and lamellar surface (cf. Fig. 2). The contact faces are again of type B (low density of methyl groups). However, due to the symmetrical orientation of the methyl groups' pattern relative to the lamellar surface, it is not possible to deduce the chain orientation from the AFM pictures, and thus, not possible to determine the hand of exposed helices (Stocker et al., to be published).

References

1. Brückner S, Meille SV, Petraccone V, Pirozzi B (1991) Progr Polym Sci 16:361—404
2. Brückner S, Meille SV (1989) Nature 340:455
3. Lotz B, Fillon B, Thierry A, Wittmann JC (1991) Polym Bull 25:101
4. Lotz B, Graff S, Straupe C, Wittmann JC (1991) Polymer 32:2902
5. Lotz B, Wittmann JC, Stocker W, Magonov SN, Cantow HJ (1991) Polym Bull 26:209
6. Wittmann JC, Lotz B (1990) Progr Polym Sci 15:900
7. Fillon B (1989) Thesis, University of Strasbourg
8. Varga J, Toth F (1986) Makromol Chem Makromol Symp 5:213
9. Icenogle RD (1985) J Polym Sci, Polym Phys Ed 23:1369
10. Varga J (1986) J Therm Anal 165:31
11. Lotz B, Wittmann JC (1986) J Polym Sci, Polym Phys Ed 24:1541
12. Khoury F (1966) J Res Natl Bur Stand 70A:29
13. Lotz B, Graff S, Wittmann JC (1986) J Polym Sci, Polym Phys Ed 24:2017
14. Stocker W, Bar G, Kunz M, Möller M, Magonov SN, Cantow HJ (1991) Polym Bull 26:215

Authors' address:

Bernard Lotz, Jean Claude Wittmann
Institut Charles Sadron (CRM-EAHP), CNRS-ULP
6, rue Boussingault
67083 Strasbourg Cédex, France

Crystallization and morphology of binary blends of linear and branched polyethylene

U. W. Gedde

Department of Polymer Technology, Royal Institute of Technology, Stockholm, Sweden

Abstract: This paper reviews crystallization kinetics and morphology studies of binary blends of a low molar mass linear polyethylene sharp fraction (M_w = 2500) and higher molar mass samples of various linear and branched polyethylenes. Cocrystallization and segregation phenomena occurring in these polymers are reported as a function of molar mass and degree of chain branching of the high molar mass component, blend composition, and crystallization temperature. Branched polyethylene with 1.5 mol% of ethyl groups cocrystallizes with the low molar mass linear PE fraction, whereas segregation of the low molar mass polymer is the dominant feature in the blends with polymers having a lower degree of chain branching ($\leqslant$0.5 mol%). The applicability of differential scanning calorimetry, polarized light microscopy, and transmission electron microscopy to reveal cocrystallization and segregation phenomena in polyethylene is discussed.

Key words: Polyethylene; binary mixtures; crystallization; morphology; segregation

Introduction

Crystallization of most polymers is accompanied by the separation of different molecular species, a process referred to as molecular fractionation. In linear polyethylene (LPE), fractionation occurs due to differences in molar mass [1—4]. The low molar mass material crystallizes at low temperatures in subsidiary lamellae located between the dominant lamellae and in the spherulite boundaries [5, 6].

Wunderlich and Mehta [1, 2] summarized in 1974 the experimental evidence for molecular fractionation in LPE and suggested a hypothesis in which the concept molecular nucleation played a vital role. According to Wunderlich and Mehta, there exists at each crystallization temperature a critical molar mass (M_{crit}) such that the molecule of a molar mass greater than M_{crit} are able to crystallize at this temperature, whereas molecules of molar mass less than M_{crit} are unable to crystallize. Figure 1 presents the data of Wunderlich and Mehta [1, 2] who showed that M_{crit} was roughly proportional to 1/(undercooling)^2.

Most studies concerned with molecular fractionation dealt with samples having a broad continuous molar mass distribution. The crystallization of binary mixtures of sharp fractions was studied to a lesser degree. The crystallization of binary mixtures of LPE sharp fractions in the molar mass range from 1000 to 20 000 dependent upon the cooling rate, and two types of crystallization were observed [7]: i) separate crystallization of the components occurred at low degrees of undercooling, and ii) water-quenched mixtures crystallizing at very extensive degrees of undercooling displayed only one melting peak and one small-angle x-ray scattering peak which was taken as evidence for cocrystallization of the components.

Polymers with branching and tacticity exhibit not only molar mass segregation, but also segregation phenomena relating to these structural irregularities. Figure 1 shows that the crystallization temperature range is shifted towards lower temperatures with increasing chain branching [8]. The multicomponent nature of branched polyethylene (BPE) arises from the fact that the chain

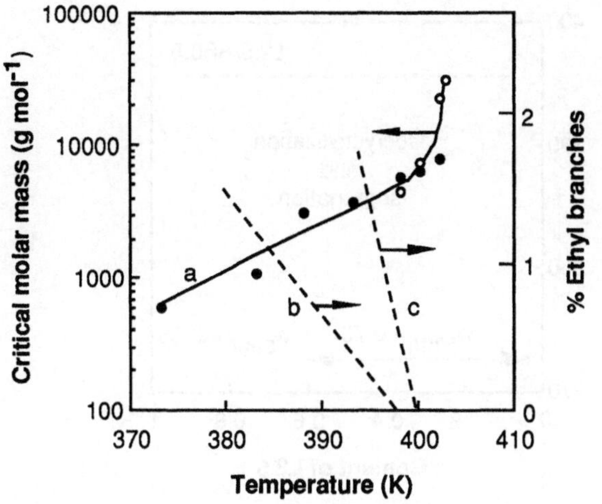

Fig. 1. Critical molar mass of LPE as a function of crystallization temperature (curve a; drawn after data by Wunderlich and Mehta [2]). Curves b and c show the crystallization temperature range as a function of the content of ethyl branches; lines b and c correspond to the temperatures at which 50% of the final crystallinity is completed within 100 s and 10 000 s, respectively. Data from [8, 21]

branches are randomly positioned on the polymer backbone chain. Segregation is thus never sharp, as is indicated in Fig. 1.

Blends of LPE and BPE have received considerable attention during recent years [9—14]. Unquestionably, the two components in binary mixtures of linear and high pressure BPE are essentially incompatible in the solid state [9-11]. Blends of LPE and low pressure BPE have been studied by several authors and their conclusions are very different, although their studies have concerned similar polymers of relatively high molar mass with medium to high polydispersity and with the branched polymer containing 1.4—1.8 mol% of ethyl groups [12—14]. Hu et al. [12] and Edwards [13] presented evidence by differential scanning calorimetry (DSC), x-ray diffraction, and Raman spectroscopy in support of cocrystallization of LPE and BPE in slowly cooled samples.

Contrary to this view, Norton and Keller [14] reported data obtained by DSC, polarized light microscopy, and transmission electron microscopy (TEM) establishing, predominantly, segregation of LPE and BPE (1.4 mol% of ethyl groups) components in a 50/50 blend of commercial HDPE and LLDPE crystallized at different constant temperatures between 394 and 403 K. The linear polymer crystallized first under isothermal conditions, forming regular-shaped sheets with a spherulitic morphology, whereas the branched polymer crystallized at a later stage (during the rapid cooling) in finer, S-shaped lamellae located between the stacks of dominant lamellae. Some, limited co-crystallization was, however, indicated in samples crystallized close to 394 K. According to DSC, the quenched samples exhibited less pronounced segregation.

This paper reviews the work performed on binary blends of a low molar mass LPE (M_w = 2500; M_w/M_n = 1.15; denoted L2.5) and higher molar mass polymers of LPE and BPE [15—22]. The topics discussed are crystallization kinetics, whether the components cocrystallize/segregate, the lamellar and spherulitic morphology, and the methodology to reveal cocrystallization/segregation of binary blends.

Results and discussion

Binary blends of LPE fractions of different molar mass

Crystallization and morphology were studied in binary blends of the low molar mass LPE fraction (L2.5) and several higher molar mass sharp fractions: M_w = 11 000 (L11), 22 000 (L22), and 66 000 (L66) [15—17].

Figure 2 illustrates the different types of crystallization observed in the linear PE blends [16]: a) At high temperatures the high molar mass

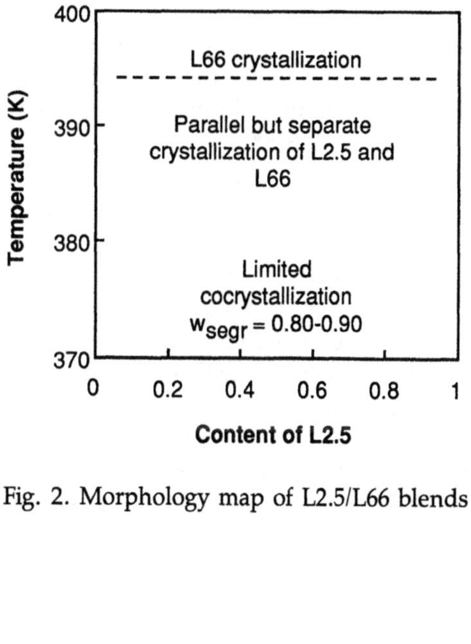

Fig. 2. Morphology map of L2.5/L66 blends

polymer crystallizes alone. Data for the fold surface free energy obtained from linear growth rate data supported the view that the nature of the fold surface of the dominant lamellae is related only to the molar mass of the crystallizing component and is not affected by the composition of the melt; b) At intermediate temperatures, i.e., at temperatures below the temperature corresponding to M_{crit} = 2500, both components crystallize, but in different crystal lamellae. Crystallization of the low molar mass component in the blend is promoted by the presence of crystals consisting of the high molar mass material. c) At low temperatures, partial cocrystallization was suggested based on TEM and DSC data. Binary mixtures L2.5/L66 being crystallized at temperatures from 380 to 403 K displayed, depending on temperature, three types of morphology [15]: Axialites were formed at the highest temperatures, non-banded spherulites at intermediate temperatures, and banded spherulites at the lowest temperatures. In the composition range 0.2 to 1.0 w/w L66, the samples studied, showed a distinct axialite-spherulite transition at a temperature of 16.5 K to 17.5 K below the equilibrium melting point of L66. At these temperatures, L66 crystallizes alone and L2.5 acts as diluent which merely has the effect of reducing the equilibrium melting point of the crystallizing L66. The axialite-spherulite transition was accompanied by the commonly observed doubling of the kinetic constant (K_g) [16].

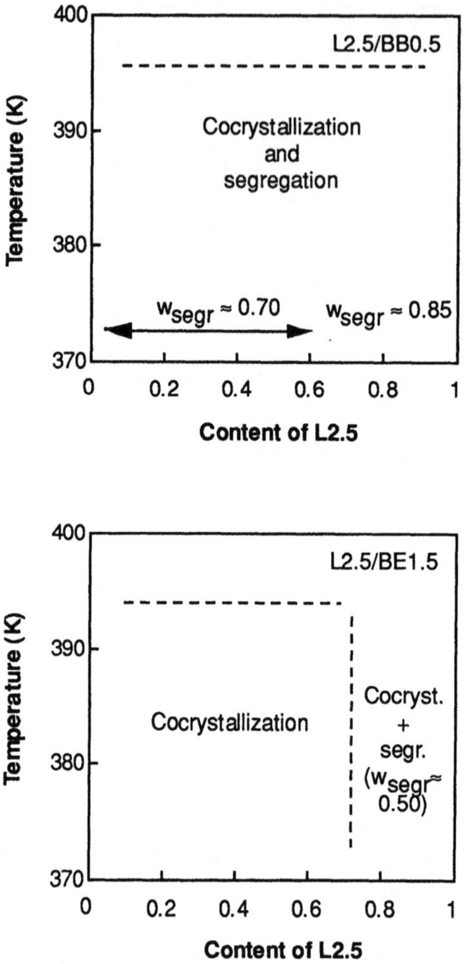

Fig. 3. Morphology map of LPE/BPE blends. Upper graph: L2.5/BB0.5; bottom graph: L2.5/BE1.5

Binary blends of LPE and BPE

Crystallization and morphology were studied in binary blends of the low molar mass fraction L2.5 and two higher molar mass branched polymers: sample BE1.5: M_w = 166 000; M_w/M_n = 6.1; 1.5 mol% of ethyl branches; sample BB0.5: M_w = 290 000; M_w/M_n = 14.5; 0.5 mol% of butyl branches [18–22].

In binary blends of L2.5 and branched PE (BE1.5 and BB0.5), TEM showed a progressive change from curved to straight lamellae and a strong decrease in average amorphous thickness with increasing content of the linear polymer [19, 22]. Data obtained by model calculations of the average amorphous thickness assuming complete co-crystallization of the linear and branched polymers show good correspondence with the experimental data obtained for blends of L2.5 and BE 1.5, except for the blend consisting of 80% of L2.5, but show a pronounced deviation for blends of L2.5 and BB0.5. For most of the blends of L2.5 and BE1.5 there was good agreement between calorimetric crystallinity and crystallinity determined by TEM, indicating cocrystallization of the components. There was, however, a significant deviation between the two crystallinity values for all blends of L2.5 and BB0.5, and for the blend of L2.5 and BE1.5 with 80% of L2.5, which can be explained by partial segregation of L2.5. The linear growth rate and the supermolecular structure were found to be highly sensitive to composition [21]. The pronounced increase in linear growth rate with increasing content of the linear fraction may be explained by an increase in the rate of diffusion of crystallizable segments due to a reduction in chain entanglement. The introduction of the linear polyethylene fraction changed the originally spherulitic structure into a predominantly axialitic one.

Barham et al. [23] presented evidence in favor of a liquid-liquid segregation in mixtures of LPE and BPE and such pre-crystallization segregation may in fact be operative in the L2.5/BE1.5 (0.8/0.2) blend.

Methodology to reveal segregation and cocrystallization in PE blends

DSC is here a very useful method. It is easy and informative to count the number of melting peaks. Isothermal crystallization followed by quenching normally results in a sample which exhibits two melting peaks: the high temperature peak is due to the material crystallizing under isothermal conditions, whereas the low temperature peak is due to the segregated species crystallzing during quenching. Segregation of the different components in binary blends first isothermally crystallized for a certain period of time and then rapidly cooled to room temperature should be revealed by the presence of two, three, or perhaps four melting peaks. Complete and solitary crystallization of the readily crystallizing component under the isothermal conditions followed by crystallization of the second component during quenching result in a sample with only two melting peaks. Separate crystallization occurring during the isothermal phase ("isothermal segregation") is indicated by the presence of two melting peaks located at temperatures higher than the crystallization temperature. Figure 7 in [16] illustrates the danger of counting the number of melting peaks. The melting peaks of the isothermally crystallized L2.5- and L66-rich material were overlapping in the samples crystallized for only a short period of time, and segregation was only revealed upon prolonged isothermal treatment when the melting of L66-rich crystals was shifted towards higher temperatures by crystal thickening. DSC revealed isothermal segregation occurring in L2.5/BE1.5 (0.8/0.2). Two high-temperature melting peaks were observed in addition to the low-temperature peak associated with the cocrystals formed during quenching. The presence of separate crystals of L2.5 was also confirmed by TEM [22].

Polarized light microscopy was used by Norton and Keller [14]. They observed the rejection of low melting branched species towards the spherulite boundaries in mixtures of HDPE and LDPE. At high temperatures, the HDPE crystallized first in coarse, large spherulites and fine-grained material consisting essentially of segregated branched material was formed upon subsequent cooling [14]. The mixtures of L2.5 and the various LPE and BPE fractions exhibited only micro-segregation [15, 19, 22]. Mixtures of L2.5/L66 with a high L2.5 content exhibited vague spherulites of low birefringence when crystallized at very high temperatures [15]. The low concentration of L66 in these blends resulted in a low density of dominant crystal lamellae surrounded by the L2.5-rich melt. The spherulitic structure appeared with "normal" intensity after cooling due to crystallization of the L2.5-rich material present in the intra-spherulitic pockets.

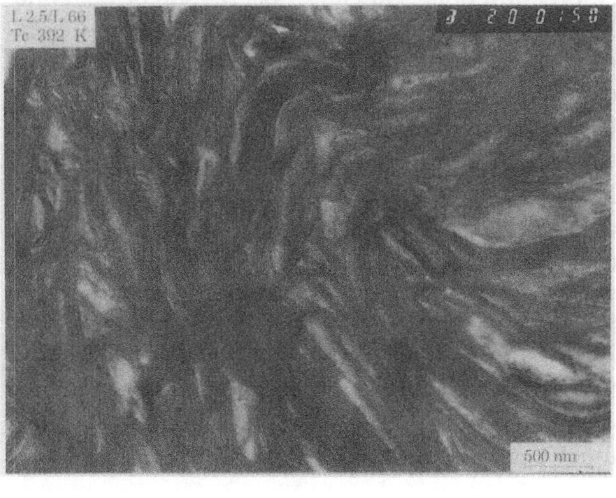

Fig. 4. Transmission electron micrograph of chlorosulphonated L2.5/L66 (0.5/0.5) crystallized at 392 K. Unstained areas of segregated low molar mass material (L2.5) are shown

Segregation of low molar mass polymer (L2.5) was, however, best revealed by TEM on chlorosulphonated thin sections. A number of independent observations can be made by studying the electron micrographs, each of them informative about the state of mixing of the systems [22]. The presence of isolated unstained areas, as shown in Fig. 4, is strong evidence for the presence of

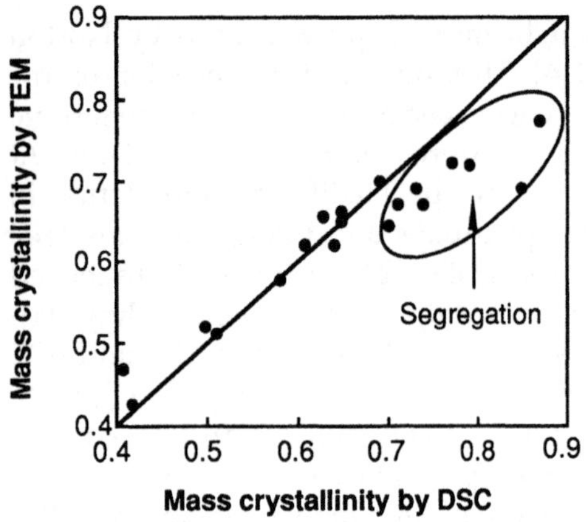

Fig. 5. Mass crystallinity obtained by TEM from crystal and amorphous thickness data plotted versus mass crystallinity by DSC according to the total enthalpy method [24]. Data for blends of L2.5/BE1.5 and L2.5/BB0.5 are included. Drawn after data in [22]

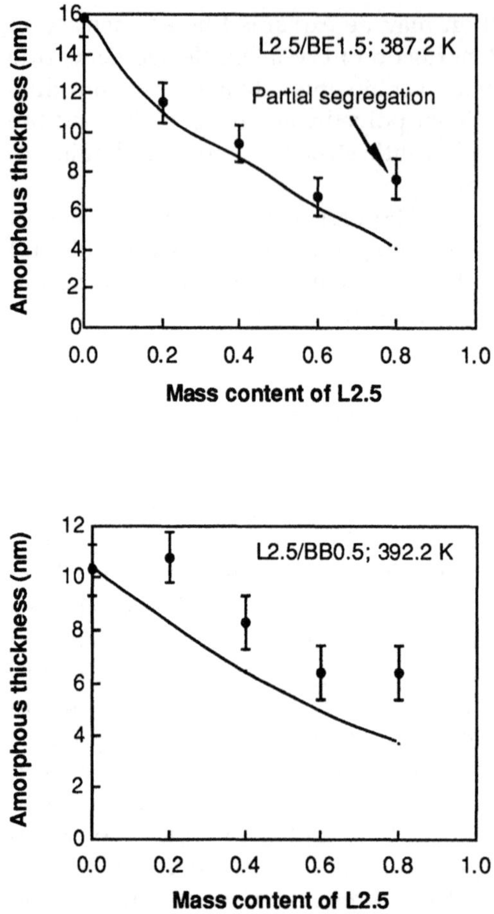

Fig. 6. Average thickness of amorphous interlayer plotted as a function of mass content of L2.5. The solid lines are obtained from model calculations according to Eq. (2): Upper diagram: L2.5/BE1.5; crystallized at 387.2 K for 0.7 h. Lower diagram: L2.5/BB0.5; crystallized at 392.2 K for 12 h. (From [22] with permission from Butterworth & Heinemann Ltd., U.K.)

segregated L2.5. The ultra-thin amorphous layers present in L2.5 could not be stained by the chlorosulphonic acid [16]. Most blends of L2.5 and L66 exhibited stained areas showing long and comparatively thick crystal lamellae surrounding areas of unstrained, low molar mass material [16, 17]. Painstaking measurement of the thickness of several hundred amorphous layers and crystal lamellae turned out to be most useful. The crystallinities derived from the amorphous thickness and crystallite thickness distributions were compared with the DSC crystallinities. Segregation of L2.5 led to an underestimation of the crystallinity by TEM and it was possible to estimate the fraction of L2.5 being segregated (w_{segr}), using data of the type presented in Fig. 5 according to the following equation:

$$w_{segr} = \frac{w_c(\text{DSC}) - w_c(\text{TEM})}{w_{L.5}[w_{cL2.5} - w_c(\text{TEM})]} \qquad (1)$$

where $w_{L.5}$ is the content of L2.5 in the blend and $w_{cL2.5}$ is the crystallinity of pure L2.5. This technique gave results consistent with other compatibility estimates obtained by TEM which focused the average amorphous thickness (see Fig. 6).

The mass average thickness of the amorphous interlayers ($\langle L_a \rangle$) sensitively reflected the state of mixing. The low molar mass material crystallized in extended chain or once-folded crystallites, whereas BPE exhibited relatively thick amorphous interlayers. Figure 6 shows data for the mass average amorphous thickness based on several hundred measurements from micrographs of the chlorosulphonated sections plotted as a function of the composition. The solid lines present in the graphs of Fig. 6 are based on calculations using the following equation (details are given in [22]), assuming that the nature of the amorphous segments of the branched polymer is not affected by the introduction of L2.5, i.e., that the probability of tight chain folding and the average volume occupied by each random amorphous segment is independent of composition:

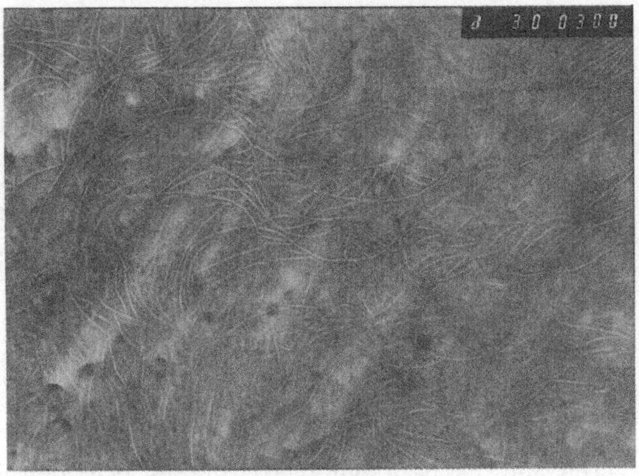

a)

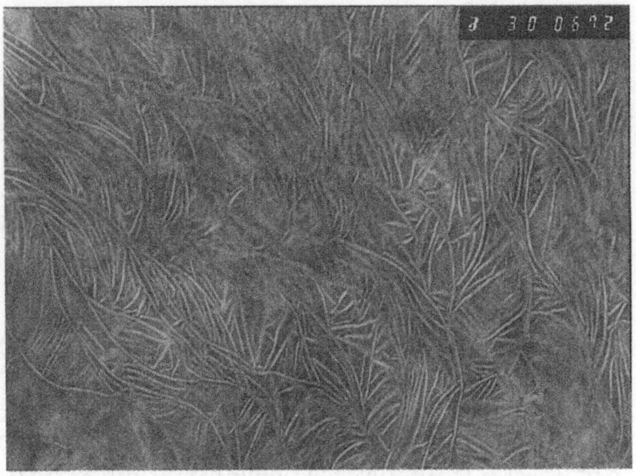

b)

c)

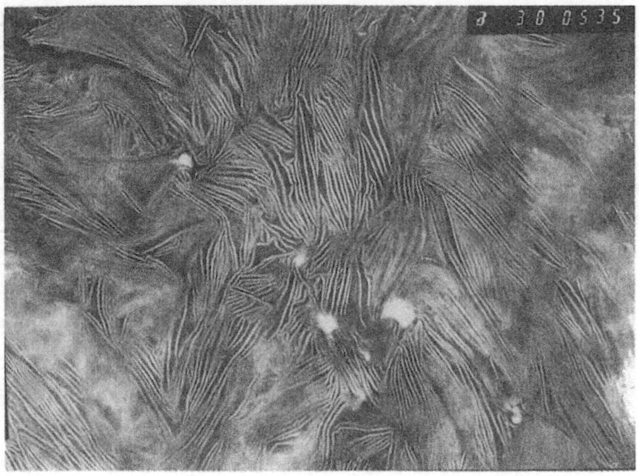

d)

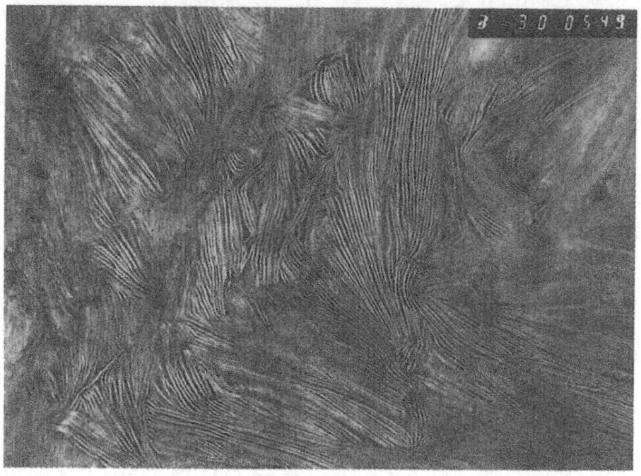

e)

Fig. 7. Transmission electron micrographs of chlorosulphonated sections of L2.5/BE1.5 crystallized at 387.2 K for 0.7 h and then cooled at a rate of 80 K min^{-1} to room temperature: a) 0% L2.5; b) 20% L2.5; c) 40% L2.5; d) 60% L2.5; e) 80% L2.5. (From [22] with permission from Butterworth & Heinemann Ltd., U.K.)

$$\langle L_a \rangle = \frac{\cos\theta}{\cos\theta_{\mathrm{BPE}}} \times \langle L_{a0} \rangle \left(\frac{[1 - w_{\mathrm{L2.5}}]w_{c\mathrm{BPE}}}{[1 - w_{\mathrm{L2.5}}]w_{c\mathrm{BPE}} + w_{\mathrm{L2.5}}w_{c\mathrm{L2.5}}} \right) + \left(\frac{2.63\cos\theta\, w_{\mathrm{L2.5}}w_{c\mathrm{L2.5}}}{[1 - w_{\mathrm{L2.5}}]w_{c\mathrm{BPE}} + w_{\mathrm{L2.5}}w_{c\mathrm{L2.5}}} \right), \quad (2)$$

where $w_{c\mathrm{BPE}}$ is the mass crystallinity of the branched polymer, $w_{c\mathrm{L2.5}}$ is the mass crystallinity of L2.5, θ is the chain tilt angle of the sample, θ_{BPE} is the chain tilt angle of the branched polymer, and $\langle L_{a0} \rangle$ is the mass average amorphous thickness of the pure branched polymer. It is also assumed that L2.5 contributes to the amorphous base by 10% of its total volume, independent of composition.

The experimental values of the L2.5/BE1.5 blends essentially follow Eq. (2), indicating cocrystallization, except for the 0.8/0.2 blend. The experimental value of the latter is greater than the predicted value which indicates partial segregation of L2.5 in this sample. Segregated L2.5 material in this sample is not accounted for in the amorphous thickness distribution for reasons given earlier. The same is true of all blends of L2.5 and BB0.5 (Fig. 6).

The shape of the crystal lamellae in the L2.5/BPE blends also exhibited a strong composition dependence, as is illustrated in Fig. 7 which shows the blends of L2.5 and BE1.5 crystallized at 387.2 K [22]. The blends with 0% and 20% of L2.5 displayed long S-shaped lamellae surrounding significantly shorter lamellae of the same or lower thickness after crystallization at 387.2 K (Fig. 7a, b). The S-shape of these lamellae was more pronounced for pure BE1.5 than for the blend with 20% of L2.5. A general feature of these samples was also the presence of large domains of only thin S-shaped short lamellae. A substantial change in morphology occurred in the case of the blend consisting of 40% of L2.5 (Fig. 7c). The lamellae in this sample were dominantly short, straight or slightly curved and more occasionally roof-ridge-shaped. For the blends containing 60% or more of L2.5, the lamellae were straight and occassionally roof-ridged (Figs. 7d, e). The average width of the crystals increased with increasing L2.5 content from 0.2 μm in pure BE1.5 to 0.5 μm in the 80% L2.5 blend. The number of lamellae per stack increased substantially compared with the blends having a lower content of L2.5. A monotonous decrease in thickness of the amorphous interlayers with increasing content of L2.5 was also strikingly apparent in the micrographs, as is shown in Fig. 6. Norton and Keller [14] showed that S-shaped lamellae typical of BPE crystals remained in segregated mixtures of LDPE and HDPE.

Acknowledgements

This study has been sponsored by the Research Council of the National Swedish Board for Technical Development (STUF), grant 90-00434P. The cooperation during the course of this work of Drs. J. M. Rego Lopez, M. T. Conde Braña, J. I. Iragorri Sainz, J.-F. Jansson, and B. Terselius is gratefully acknowledged.

References

1. Wunderlich B, Mehta A (1974) J Polym Sci, Polym Phys Ed 12:255
2. Wunderlich B, Mehta A (1975) Coll Polym Sci 253:193–205
3. Gedde UV, Jannson J-F (1983) Polymer 24:1521–1531
4. Gedde UW, Edklund S, Jansson J-F (1983) Polymer 24:1532–1540
5. Gedde UW, Jansson J-F (1984) Polymer 25:1263–1267
6. Bassett DC (1986) In: Sedlácek B, "Morphology of Polymers", Proc 17th Europhysics Conf on Macromolecular Physics, Prague, Czechoslovakia, July 1985, Walter de Gruyter, Berlin and New York, pp 47–68
7. Smith P, St. John Manley R (1979) Macromolecules 12:483
8. Gedde UW, Jansson J-F, Liljenström G, Eklund S, Wang P-L, Holding S and Werner P-E (1988) Polym Eng Sci 28:1289–1303
9. Clampitt BH (1965) J Polym Sci 3:671
10. Datta NK, Birley AW (1982) Plast Rubb Process Appl 2:237
11. Kyu T, Hu S-R, Stein RS (1987) J Polym Sci, Polym Phys Ed 25:89
12. Hu S-R, Kyu T, Stein RS (1987) J Polym Sci, Polym Phys Ed 25:71
13. Edwards GH (1986) Brit Polym J 18:88
14. Norton DR, Keller A (1984) J Mater Sci 19:447
15. Rego Lopez JM, Gedde UW (1988) Polymer 29:1037–1044
16. Rego Lopez JM, Conde Braña MT, Terselius B, Gedde UW (1988) Polymer 29:1045–1051
17. Conde Braña MT, Iragorri Sainz JI, Gedde UW (1989) Polym Bulletin 22:277–284
18. Rego Lopez JM, Gedde UW (1989) Polymer 30:22–26

19. Conde Braña MT, Irgaorri Sainz JI, Terselius B, Gedde UW (1989) Polymer 30:410—415
20. Gustafsson A, Conde Braña MT, Gedde UW (1991) Polymer 32:426—431
21. Iragorri Sainz JI, Rego Lopez JM, Katime I, Conde Braña MT, Gedde UW (1992) Polymer 33:461—466
22. Conde Braña MT, Gedde UW (1991) Polymer (1992) 33:3123—3136
23. Barham PJ, Hill MJ, Keller A, Rosney CCA (1988) J Mater Sci Lett 7:1271—1275
24. Gray AP (1970) Thermochim Acta 1:563

Author's address:

Dr. Ulf W. Gedde
Department of Polymer Technology
Royal Institute of Technology
S-10044 Stockholm, Sweden

Progress in Colloid & Polymer Science Progr Colloid Polym Sci 87:16—24 (1992)

Electron microscopic investigation of the crystallization behavior of polymer blends

G. H. Michler[1]), I. Naumann[1]), and H. Steinbach[2])

[1]) Technical University Merseburg, Institute of Applied Physics, Merseburg, FRG
[2]) BUNA AG ZB Research & Development, Schkopau, FRG

Abstract: Polymer blends containing semicrystalline components possess a complex morphology, including the distribution of the polymer components and the details of the semicrystalline structure. Electron microscopy was used to reveal the morphology and the crystallization behavior of several polymer blends. In blends of SAN and polyurethane (PU) there is an increased affinity of the PU to SAN, yielding lamellae of PU-like needles in the SAN matrix. A modification of blow foils of high-density polyethylene (HDPE) by small amounts of a rubber improves the packing of lamellae parallel to the normal direction of the foil (improved c-texture). Blends containing linear (HDPE), branched (LDPE), and linear low-density polyethylene (LLDPE) show lamellae containing only one type of macromolecule, as well as lamellae containing macromolecules of different PE components (cocrystallization). Detailed information has been obtained by quantitative measuring of thickness distributions of lamellae and by determination of lamellae in the melting range after fixation of lamellae using γ-irradiation.

Key words: Electron microscopy; semi-crystalline polymers; polymer blends; cocrystallization; thickness distribution of lamellae

Introduction

Polymer blends are widely used in many fields of applications to realize special properties or property combinations. The properties achieved strongly depend on the distribution of the polymer components in the blend, i.e., on the morphology. Usually, the polymer components are incompatible, forming separate phases with more or less pronounced boundary layers between them. Besides the distribution of the polymer components, the internal structure of each of the polymer phases is also important. If one or more components consists of a semi-crystalline polymer, the distribution of the crystalline elements and the amorphous phase should be considered. In these semi-crystalline polymer blends a complex morphology with structural details in many different levels of size exists [1]. Of particular interest is the question of whether and to what degree the conditions of crystallization during the solidification process are changed by the presence of a second polymer component.

From the view of morphology three types of blends with semicrystalline components can be distinguished:

i) The polymer components clearly form separate phases without or with only very weak interactions at the phase boundaries.
ii) There are separate polymer phases, but with some phase interactions.
iii) The polymer components do not show clearly pronounced boundaries between them. The crystallization of one component appears in close proximity to the other component(s) .

It is clear that such a classification is not absolute, but depends to a high degree on the processing conditions and, particularly, on the intensity of mixing.

Often used mixtures of semi-crystalline polymers are blends of different types of polyethylenes, i.e., of high-density (HDPE), low-density (LDPE), and linear, low-density polyethylene (LLDPE). In the literature HDPE/LDPE blends have been described as incompatible [2—4] or (in some composition ratios) as compatible blends [5—7]. Compatibility means that different types of macromolecules can crystallize in the same crystals or lamellae. Cocrystallization could be possible from a crystallographic point of view, because of the somewhat small differences of the lattice parameters between HDPE and LDPE. But other results also show in the pure components HDPE or LDPE a demixing of macromolecules in dependence on molecular weight (e.g., in extended chain crystals [8], in paraffins [5], and in blends of fractions of PE [7a]).

One reason for this controversion is the strong influence of the processing conditions, including intensity of mixing and cooling rate [6, 7].

Recently, effects of partial demixing, not only after cooling, but also in the melt (i.e., a liquid-liquid phase segreation) have been described in blends of the type HDPE/LDPE [9], HDPE/UHMWPE, and LDPE/LLDPE [10]. A thermodynamical demixing of macromolecules of very low density PE (VLDPE) which differs in branching content is discussed and correlated with a domain-like morphology in [11].

To study the distribution of the polymer components as well as the crystallization behavior of the semi-crystalline polymers the techniques of electron microscopy are very helpful. Particularly, the quantitative evaluation of distribution of thickness of lamellae [12] as well as a special technique to study lamellae directly in the melting range after electron- or γ-irradiation [13—15] have been applied to study different semi-crystalline polymer blends. Some results are presented and discussed in this work.

Samples

The following mechanical mixtures have been used:

- blends of segmented polyurethanes (PU) and acrylonitrile-styrene-butadiene-terpolymers (ABS): PU is a thermoplastic polyesterurethane based on MDI, with a glass transition temperature of T_g = —24 °C. ABS is a commercial material of the BUNA AG (Sconater 442 MA). Samples were mixed in a double-screw extruder in a wide range of composition.
- blends of segmented polyurethane (PU) and acrylonitrile-styrene-copolymer (SAN): SAN is a commercial material (Sconarol, BUNA AG); composition SAN/PU = 80/20.
- blends of HDPE (ρ = 0.951 g/cm^3, $\bar{M}_w$ = 180 000, $U = \bar{M}_w / \bar{M}_n$ = 10.3) and rubber (composition of styrene-butadiene-rubber and natural rubber) with rubber contents up to 10%. Samples were mixed in a single screw extruder and foils about 60—80-µm thick were prepared in a film-blowing unit (drawing ratio 1:8.5, blowup ratio 1:1.4 [19].
- blends of branched PE (LDPE) and linear PE (HDPE) in the whole composition range:

  LDPE: ρ = 0.9204 g/cm^3, M_w = 330 000, U = 16, branching degree 2.0 CH_3/100 C;
  HDPE: ρ = 0.9520 g/cm^3, M_w = 140 000, U = 5.7, branching degree <0.3 CH_3/100 C.
- blends of 70% branched PE and 30% linear low-density PE (LLDPE).

After mixing in the melt, injection-molded samples or pressure-molded plates were prepared (besides the blown films) and slowly cooled (with a cooling rate of about 1 K/min).

Investigation techniques

To study the distribution of the polymer components (phase morphology) and the internal structure of each phase (semicrystalline morphology) the usual technique of ultramicrotomy and transmission electron microscopy (TEM) has been used. To enhance the contrast between the structural details the samples were treated by selective chemical-staining procedures using chlorosulfonic acid and osmium tetroxide (for PE), and a combination of chlorosulfonic acid, osmium tetroxide and formalin (for PU).

To study the melting behavior (or crystallization) a special in situ technique has been used [15]: The samples were heated to different temperatures in the melting range of the polymer components investigated and γ-irradiated at these temperatures. Up to doses of more than 10 MGy, γ-irradiation causes cross-linking mainly in the amorphous regions, whereas the crystalline parts (i.e., the lamellae) are essentially unmodified [13]. During subsequent cooling the amorphous parts are fixed in their structure and are unable to crystallize. Only those lamellae which exist at the irradiation temperature are present in the sample after cooling. After selective chemical staining and cutting ultrathin sections these lamellae are studied in TEM.

In some cases the distribution of the thickness of lamellae has been quantitatively determined. On electron micrographs equidistant lines were drawn and at each crossing point with lamellae the corresponding thicknesses were measured. The number of lamellae per μm^2 were also determined directly on micrographs.

Results and discussion

Blends with clearly separated phases

a) PU/ABS-, PU/SAN-Blends

Blends of PU and ABS show a phase separation between the two components over the whole range of composition. But there is a peculiarity at the phase boundaries with only a contact of PU to the SAN-phase of ABS and not to the rubber phase (polybutadiene), i.e., there is an increased affinity of PU to SAN [16] (Fig. 1). This correlate with spectroscopic results on a chemical interaction between urethanamid groups of PU and nitrile groups of SAN in ABS [17].

Blends of PU and SAN also show a phase separation [18]. In the blend of Fig. 2 with 20% PU in a matrix of SAN the PU phase forms particles up to a few µm in size and with a very irregular shape. Partly, lamellae of PU penetrate into the SAN matrix, forming needle-like particles. This is a consequence of the above-mentioned interaction between PU and SAN.

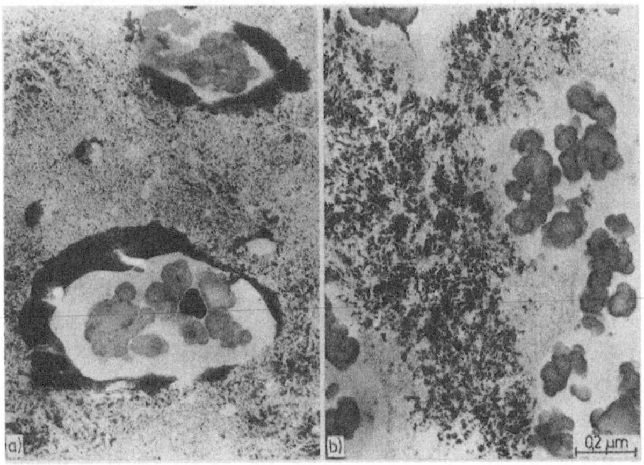

Fig. 1. Morphology of a PU/ABS-blend (with 20% ABS); ABS particles (SAN matrix with small rubber particles) in the semi-crystalline PU matrix

b) Polyethylene/rubber-blends

Usually, blow-foils of HDPE show the existence of small stacks of parallel lamellae (length up to 0.5 µm) and many randomly distributed lamellae. In accordance with this, texture analysis using wide angle x-ray scattering does not show any clear pronounced texture [19, 20]. After adding a small amount of a rubber modificator (a blend of styrene-butadiene rubber and natural rubber [19]) the blow films show a marked change in morphology and texture. There is a dominant existence of bands of parallel lamellae (Fig. 3). The bands are oriented with their length directions parallel to the machine direction MR (extrusion direction) with an angle variation of about ±30°. Since the lamellae in these bands are oriented nearly parallel to the normal direction NR, there is a clearly pronounced orientation of the stretched segments of macromolecules parallel to MR, i.e., a pronounced c-texture. This result agrees with an improved c-texture content in WAXS-measurements [19]. The rubbery modificator itself forms particles, randomly dispersed in the HDPE matrix. As a result of orientation process during extrusion these particles are oriented and elongated in machine directions with thicknesses between 0.05 and 0.5 µm and lengths between 0.1 and 10 µm. There are no sharp boundaries between rubber particles and the HDPE-matrix; the boundary fluctuates with many lamellae starting directly at the particles.

The bands of parallel lamellae appear most markedly near rubber particles. In the bands the thickness of lamellae as well as the long period (center-to-center distance of adjacent lamellae) have been measured. The corresponding frequency distribution are shown in Fig. 4 for unmodified HDPE (Fig. 4a) and for the HDPE foil modified by 10 wt.-% rubber (Fig. 4b). Whereas the frequency distributions of the thicknesses of lamellae are nearly the same in both cases, the frequency distributions of the long periods are very different. In the unmodified HDPE-foils the distribution of the long periods is broad, with a great variety of larger values. This represents the random arrangements of lamellae with interlamellar amorphous regions of different thicknesses. The modified foil shows a smaller average value of the long period, corresponding to the more densely packed lamellae in the bands. The decreased thickness of the amorphous layers indicates a higher degree of crystallinity. A measure of the local crystallinity is the relation a

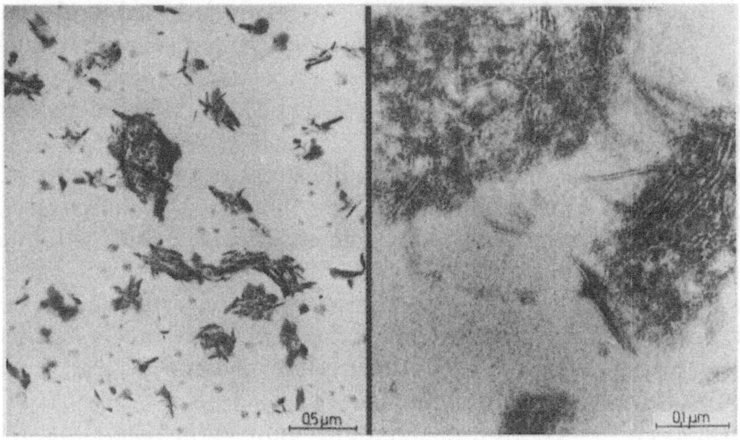

Fig. 2. Crystalline lamellae of PU grown into the SAN matrix in a SAN/PU blend with 20% PU (selectively stained ultrathin section, TEM)

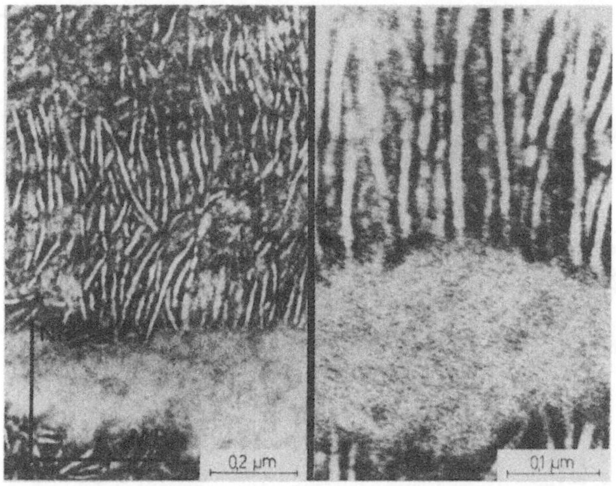

Fig. 3. Arrangement of lamellae in band-like structures in a HDPE-foil modified by 10% rubber

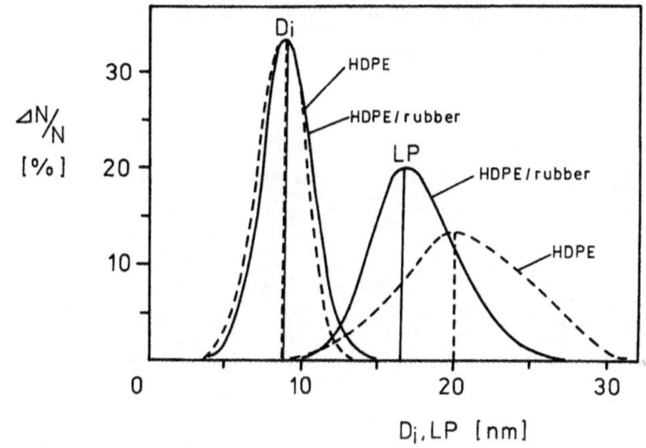

Fig. 4. Frequency distribution of the thickness of lamellae (Di) and the long period (L): a) HDPE foil; b) foil of HDPE with 10% rubber (determined from electron micrographs)

between thicknesses D_i of lamellae and long period L ($\alpha = D_i/L$). A corresponding measure of D_i and L yields an average value of $\alpha = 0.46$ for unmodified HDPE and of $\alpha = 0.54$ for modified HDPE. This result agrees with the increase in crystallinity determined by WAXS [19]. Directly inside the bands of parallel lamellae the average long period is about 14 nm (at an average thickness of lamellae of 9 nm). Therefore, the interlamellar, amorphous regions are relatively thin (about 5 nm in average) compared to the average thickness of the amorphous layers of about 10 nm in unmodified HDPE. These thin amorphous layers, together with the high degree of orientation of the macromolecular segments in the lamellae parallel to MR (c-texture), leads us to the assumption that the order of macromolecules in these amorphous regions is somewhat improved (e.g., by a higher degree of orientation, a larger content of tie-molecules or stretched entanglements).

Maximum orientation effect by rubber modification is reached between a rubber content of between 6 and 10 wt.-% [19]. The modification of HDPE appears in the melt, enabling an easier orientation of macromolecules parallel to stresses in extrusion direction. The increase of molecular mobility by the rubber modificator has also been shown by a changed deformation behavior of these foils [21].

Blends of different types of polyethylenes

a) LDPE/HDPE-blends

In these blends the largest structural details are stacks of parallel lamellae with lengths between 1 and 3 μm and thicknesses up to 1 μm. Besides these stacks the lamellae are randomly distributed with lengths between 0.03 and 1 μm; maximum length of lamellae up to more than 2 μm appears in blends with a HDPE content of 30 to 70%.

Thicknesses of lamellae depends clearly on blend composition (Fig. 5). LDPE shows thin lamellae with a relatively narrow thickness distribution, whereas in HDPE thicker lamellae with a broader distribution dominate. In the blends a coexistence of long, thick lamellae and of short, thin lamellae is visible (Fig. 6). Frequency distributions of thicknesses of lamellae determined directly by electron micrographs are shown in Fig. 7. Frequency distribution of the blends are relatively broad, and comparable with that of HDPE. The maximum as well as the average thickness of lamellae increase continuously with increasing HDPE-content, except for the blend 50/50.

There are also modifications of the internal structure of lamellae. Inside the lamellae, defect layers about 3-nm thick exist separating crystalline blocks (micro-blocks). With increasing HDPE content in the blends the number N of the defect layers per μm^2 decrease and the size B of the crystalline blocks (in length direction of lamellae, i.e., in b-direction) increase.

Electron micrographs show immediately that the two PE-types do not form separate phases. But it is not readily apparent if there is a mixing of lamellae consisting only of linear macromolecules (of HDPE-type) and of branched macromolecules (of LDPE-type) or also a mixing on molecular level with cocrystallization (joint incorporation of both types

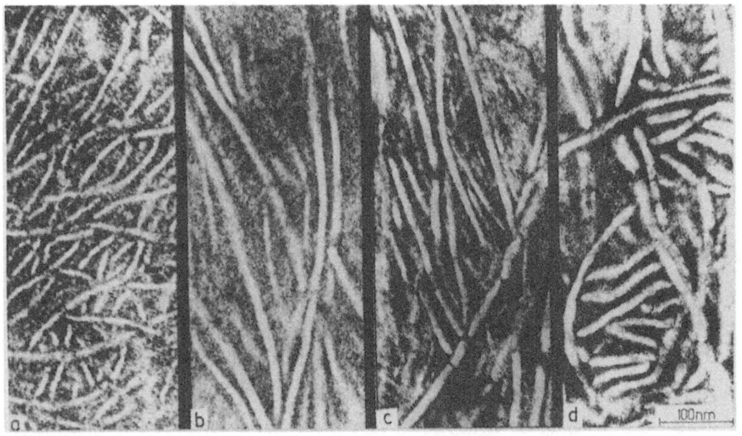

Fig. 5. Structure and size of lamellae in dependence on composition of LDPE/HDPE-blends. a) 100/0; b) 70/30; c) 30/70; d) 0/100

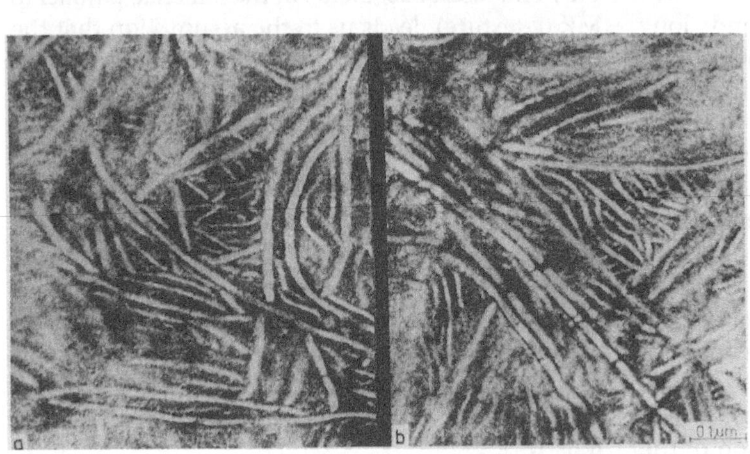

Fig. 6. Coexistence of longer thick lamellae and shorter thin lamellae as a result of a step-wise crystallization in blends of LDPE/HDPE with a composition of a) 40/60; b) 30/70

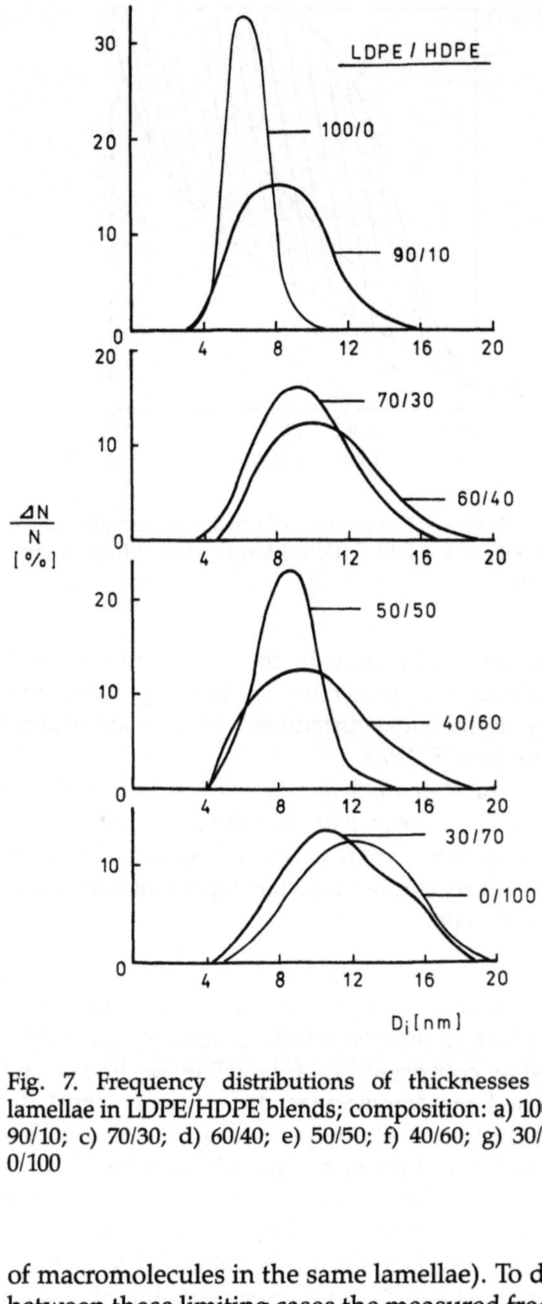

Fig. 7. Frequency distributions of thicknesses D_i of lamellae in LDPE/HDPE blends; composition: a) 100/0; b) 90/10; c) 70/30; d) 60/40; e) 50/50; f) 40/60; g) 30/70; h) 0/100

of macromolecules in the same lamellae). To decide between these limiting cases the measured frequency distributions of thicknesses of lamellae (Fig. 7) have been compared with calculated model distributions for both the limiting cases:

Case 1: with separate lamellae of the HDPE and of the LDPE-type: the expected frequency distribution of a blend is a superposition of the thickness distribution of HDPE and LDPE with a relative content corresponding to the blend composition ("additive distribution").

Case 2: with molecular miscibility in the lamellae: the expected frequency distribution of lamellae of a blend is calculated by shifting the distribution curve of LDPE towards the distribution curve of HDPE to a degree in accordance with the relative content of HDPE ("mixture distribution").

To calculate the model distributions the measured thickness distributions of lamellae of LDPE and HDPE have been used. Details of the calculation are discussed in [22]. Comparison of the calculated model distribution with the measured real distribution of the blends reveals (see Fig. 8):

- The additive model distributions show two maxima (corresponding to LDPE and HDPE). There are similarities to the real distributions only in the range of thicker lamellae.
- The mixture model distributions show one maximum. There are similarities to the real distributions only in the range of thinner lamellae and in the size of the maximum of the frequency distributions.

This reveals the following crystallization behavior during cooling from the melt:

- At higher temperatures within the crystallization range (in the range of crystallization of thicker lamellae in pure HDPE) thicker lamellae of linear macromolecules appear (lamellae of HDPE-type).
- At lower temperatures in the range between crystallization temperatures of the pure components lamellae crystallize which contain linear and only weekly branched macromolecules.
- At the lowest crystallization temperature only very thin lamellae of highly branched macromolecules appear.

This step-wise crystallization yields an arrangement of long and thick lamellae, shorter and thinner lamellae with very short and thin lamellae between them, as shown in Fig. 6. This step-wise crystallization (formation of lamellae with step-wise thinner lamellae) correlates with the crystallization model in [23].

There are two other reasons for the mechanism described above:

- In the blends the superlamellar structure, e.g., the existence of lamellar stacks, is similar to HDPE. At the beginning of crystallization the linear macromolecules form lamellae with an arrangement typical of HDPE. After that, lamellae

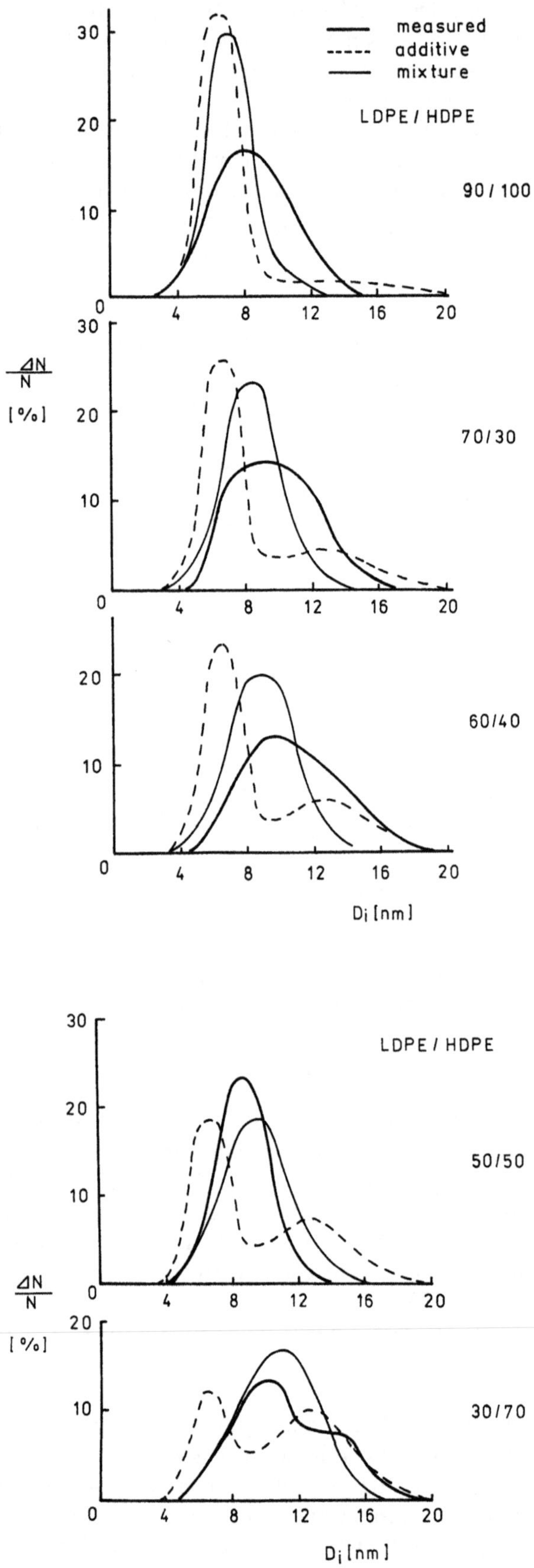

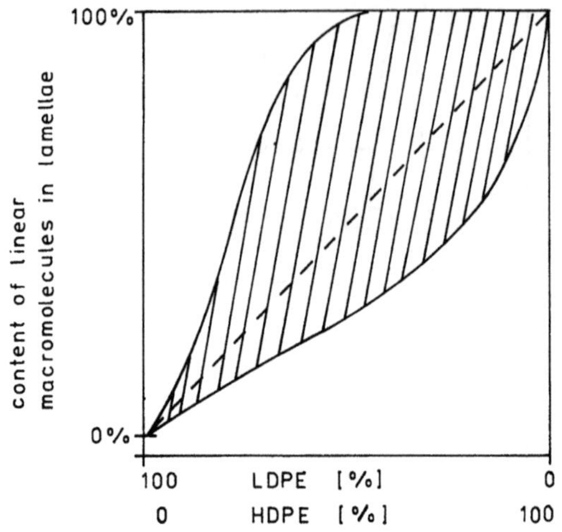

Fig. 9. Scheme of the content of linear macromolecules in the lamellae of LDPE/HDPE blends with different composition

can crystallize only in the spaces in between, without the possibility of forming their own superlamellar structures (e.g., spherulites typical of LDPE).

– The number of defect layers inside lamellae is larger and the size of microblocks smaller in the thin lamellae than in the thick, long lamellae. Therefore, thinner lamellae represent lamellae of LDPE type.

These results are summarized in the scheme of Fig. 9. Over the whole range of composition from pure LDPE to pure HDPE the scheme illustrates the probable composition of lamellae of linear and branched macromolecules. For a given blend the crystallization starts by producing lamellae with a high content of linear macromolecules (lamellae in the range above the dotted line). Lamellae crystallized later posses a higher content of branched molecules (lamellae in the range below the dotted line). The dotted line represents the ideal mixing, a cocrystallization of both types of macromolecules corresponding to the composition of the blend. For the blend LDPE/HDPE = 30/70 a content of pure HDPE-lamellae of nearly 30% has been estimated [22].

Fig. 8. Comparison of the measured frequency distribution curves of thickness of lamellae with the calculated model distributions for both the cases without miscibility ("additive distributions") and with ideal miscibility ("mixture distributions")

A cocrystallization was also found in binary mixtures of low molecular weight linear PE ($\bar{M}_w$ = 2500) and branched PE ($\bar{M}_w \approx (60—146) \cdot 1000$) using TEM of stained ultra-thin sections [7b].

b) LDPE/LLDPE blends

Investigation of the blend with 70% LDPE and 30% LLDPE in the melting range at 110° and 120°C, resp., reveals a drastic decrease in the number of lamellae (Fig. 10a). At 120°C only a few long and thick lamellae are visible. The number of lamellae per μm^2 on the micrographs is about 240 at room temperature, about 105 at 110°C, and only 25 at 120°C. The corresponding distributions of the thicknesses of the lamellae are shown in Fig. 10d. It can be seen that, with increasing temperature, the thinner lamellae are molten, but not the thickest ones. This result shows directly the step-wise character of melting and, vice versa, of crystallization.

In Fig. 11 melting behavior of both the pure components LDPE and LLDPE is compared to the melting characteristic of the blend. LDPE shows at 110°C only a few lamellae, and at 120°C lamellae are no longer visible. LLDPE shows at 110°C a small decrease and at 120°C a larger decrease in the number of visible lamellae. Only the thickest lamellae in the range of thickness variation survive. Comparison of these curves with the distribution curves of the blend (Fig. 10d) reveals: the thickest lamellae in the blend are lamellae of LLDPE. Therefore, at least the thickest lamellae in the blend are pure LLDPE lamellae, whereas the thinner lamellae, crystallizing at lower temperatures, can contain both types of macromolecules. These results confirm the conclusions about a tendency of demixing macromolecules with different branching degrees.

There are further results of studying lamellae in the melting range: Because of the lower number of lamellae, these are visible without an interaction with adjacent ones. Figure 12 shows some lamellae in PE with ultrahigh molecular weight ($M_w = 1.5 \cdot 10^6$) existing at 130°C. Directly visible are the length of the lamellae and their tendency to twist at their ends. In the middle of the micrograph a lamellae is shown with its flat side ("flat-on position" com-

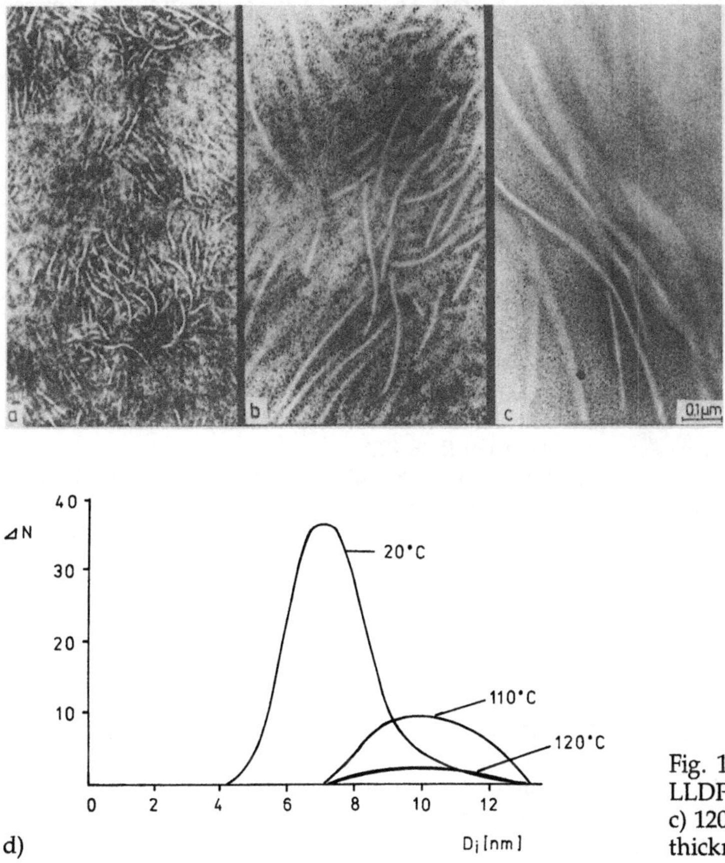

Fig. 10. Lamellae in the blend of 70% LDPE and 30% LLDPE at different temperatures: a) RT; b) 110°C; c) 120°C; d) Corresponding frequency distributions of thickness of lamellae

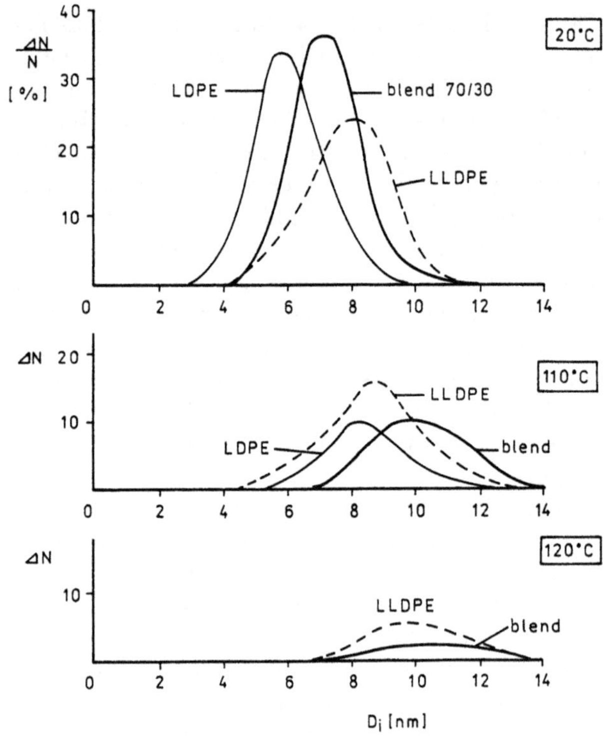

Fig. 11. Distribution of thicknesses of lamellae in LDPE, in LLDPE, and in the 70/30 blend of LDPE/LLDPE at different temperatures. a) 20°C; b) 110°C; c) 120°C

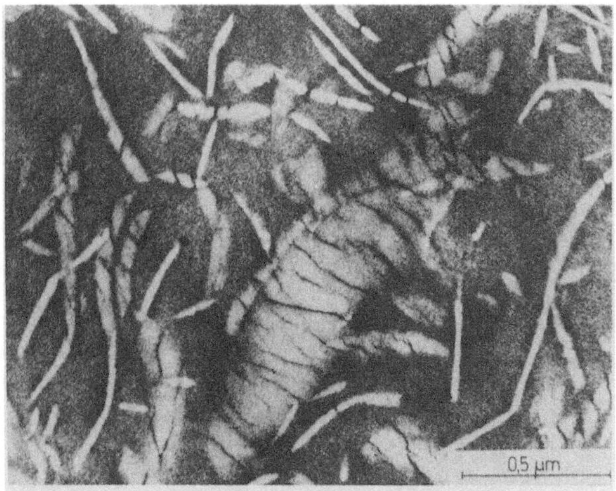

Fig. 12. Improved visibility of shape and length of single lamellae in HDPE by fixing the structure at 130°C by γ-irradiation

pared to the "edge-on position" usually visible in the electron microscope). Here, the third dimension of the lamellae (the height) is detectable. Moreover, lamellae reveal a lancett-like shape, resembling the elliptic, curved habits of lamellae after permanganic etching [24].

References

1. Martuscelli E, Palumbo R, Kryszewski M (1980) Polymer Blends: Processing, Morphology and Properties; Plenum Press New York
2. Bohn L (1966) Kolloid Z Z Polymere 213:55
3. Socher M, Häusler L, Müller U, Stephan M (1983) Plaste u Kautschuk 30:310
4. Norton DR, Keller AJ (1984) J Mater Sci 19:447
5. Asbach GI, Kilian HG (1970) Ber Bunsenges physik Chem 74:814
6. 6a. Marinow S (1982) Plaste u Kautschuk 29:202; 6b. Radusch HJ (1982) Plaste u Kautschuk 29:212; 6c. Marinow S, May M, Hoffmann K (1983) Plaste u Kautschuk 30:620, 682
7. 7a. Rego Lopez JM, Gedde UW (1988) Polymer 29:1037; 7b. Conde Brana MT, Iragorri Sainz JI, Terselius B, Gedde UW (1989) Polymer 30:410
8. Wunderlich B (1973) Macromolecular Physics Vol 1, Academic Press New York, London
9. Barham PJ, Hill MJ, Keller A, Rosney CCA (1988) J Mater Sci Lett 7:1271
10. Muller AJ et al. (1991) Proc Polymer Physics Conf Bristol P 74, P 75
11. Boode JW, Ruiten JV (1991) Proc Polymer Physics Conf, Bristol P 14
12. Michler GH, Naumann I (1982) Acta Polymerica 33:399
13. Michler GH, Gruber K, Steinbach H (1982) 33:550
14. Kanig G (1982) Colloid & Polymer Sci 260:356
15. Michler GH, Steinbach H, Bühler K (1988) Proc 12. Tagung Elektronenmikroskopie Dresden p 395
16. Michler GH, Grellmann W, Naumann I, Schierjott U, Haudel G (1989) Plaste u Kautschuk 36:333
17. Haudel G et al. (1987) Proc 31. IUPAC Macromolecular Symposium "Macro 87" Merseburg VI/Po/40
18. Radusch HJ, Hendrich R, Michler GH, Naumann I (1992) Angew Makromolek Chem 194:159
19. Marinow S, Michler GH, Walther C, Naumann I (1988) Plaste u Kautschuk 35:159; 212
20. Hofmann D, Geiß D, Janke A, Michler GH, Fiedler P (1990) J Appl Polym Sci 39:1595
21. Michler GH, Marinow S, Naumann I (1989) Plaste u Kautschuk 36:432
22. Michler GH, Steinbach H, Hoffmann K (1986) Acta Polymerica 37:289
23. Strobl GR, Engelke T, Meier H, Urban G (1982) Colloid & Polymer Sci 260:394
24. Bassett DC (1989) Proc 9. Tagung Polymerphysik Potsdam p 22; (1991) Proc Polymer Physics Conference Bristol

Authors' address:

Prof. Dr. sc. nat. G. H. Michler
Technical University Merseburg
Institute of Applied Physics
Geusaer Straße
D-O-4200 Merseburg, FRG

Role of adsorption process during crystallization of polymers at low supercooling

J. Rault

Physique des Solides, Bât. 510, Université de Paris-Sud, France

Abstract: During the process of polymer crystallization from dilute solution, the polymer chains are adsorbed on the crystal growth faces and their configurations prior to crystallization are then changed. Following the De Gennes analysis it is shown that this process leads to segregation effect; chains shorter than M^* do not crystallize. The critical mass M^* varies with the supercooling ΔT as $M^* = \Delta T^{-2}$, as was found experimentally by Wunderlich.

Key words: Polymer crystallization; adsorption; segregation

Introduction

In melt crystallization of polymers at high supercooling it has been shown [1—5] that the liquid and solid phases are correlated. The correlation law:

$$L \sim r$$

between the long period L of the semi-crystalline state and the dimension of the coils r in the melt state explains the dependence of the long period with:

a) the molecular weight distribution [1, 2];
b) the degree of branching N_b [3];
c) the melt temperature T_m [4].

The long period is the sum of the crystalline layer thickness l_c (independent of the molecular weight) and the amorphous layer thickness l_a (dependent on M). In slow cooled materials l_a is given by the relation:

$$l_a \sim \alpha \sqrt{N}\ ,$$

where N is the number of monomer units in the chain and α is a parameter which decreases when the cooling rate decreases (or when the supercooling ΔT in isothermal crystallization decreases). The origin of the amorphous phase discussed in [1, 2] is due mainly to the entanglements of the chains prior to crystallization, that is to say to the degree of overlapping of the random coils in the liquid state.

During crystallization from semi-dilute solution the chains have time to desentangle prior to (or during) crystallization ($\alpha = 0$), the amorphous phase is negligible and the chains have some regular folding. As in melt crystallization the crystalline lamellae thickness is independent of the molecular weight. In this type of crystallization, we can say that there is no correlation between the conformations of the chains before and after crystallization.

However, the aim of this work is to show that, after solution crystallization, there is a certain relationship between the solid and liquid states; the segregation of chains observed during crystallization would be due to the segregation of the liquid coils during adsorption prior to the crystallization process. The process of adsorption described hereafter and its influence on the crystallization of polymers is different than that given by Hoffmann et al. [6, 7].

Model of crystallization

Let be an isolated chain approaching, by diffusion, the growth face of a polymer crystal. A contact of one monomer of length a with the surface leads to an attractive interaction δ of the order of:

$$\delta / kT = \Delta H \Delta T / T a^3\ ,$$

where ΔH, ΔT, and T have their usual meanings [6—10].

It is to be noted that δ is of the order of 10^{-3} KT for supercooling ΔT of a few degrees. This value is obviously overestimated; the gain energy during the process should take into account the surface energies and the fact that, at the surface of the growing crystal, distortion and defects are present. Therefore, the zippering model of crystallization should be rejected (e.g., a crystalline stem of 100 carbon units of PE is not stable for a supercooling of 10 °C). The chain having several contacts with the growth face must be considered as an absorbed chain with weak interaction with surface, but with a conformation totally different from that of the random isotropic oil. During this process the chain has increasingly less translational mobility and then crystallizes by either primarily or secondary nucleation (Fig. 1).

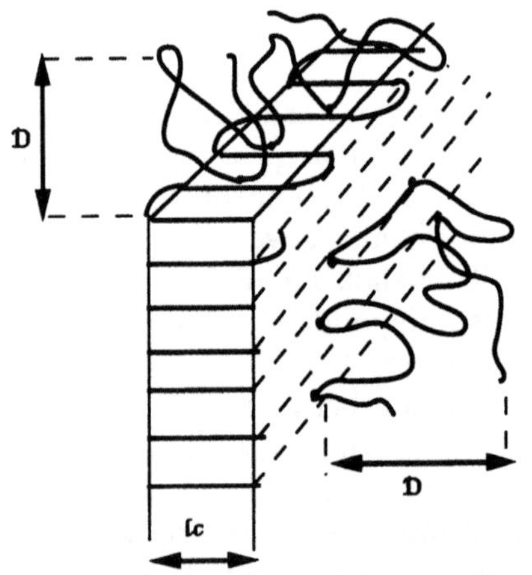

Fig. 1. Adsorption of a polymer chain on a growth face (a) and a lateral face (b) of a monocrystal; the thickness D of the adsorbed layer is independent of the molecular weight [9]; only long chains are absorbed; they then can crystallize by secondary nucleation (a) and by primary and intramolecular nucleation (b)

The central assumption in the present work is that the crystallization of absorbed chains on a growth face is slow enough for the adsorbed chains to reach their equilibrium conformation. However, for reasons of simplification and clarity, one assumes that the process of crystallization is sufficiently rapid to avoid the formation of concentrated layer of absorbed chains; De Gennes [11, 12] has discussed in detail the adsorption of polymer in dilute and semi-dilute solutions. Only the regime of dilute adsorbed chain (far from the plateau regime) is recalled here.

Let D be the thickness of the adsorbed layer; the number of contacts with the growth face is $N\,a/D$; the free energy F per chain in a θ solvent is:

$$F/KT = \frac{Na^2}{D^2} - \frac{\delta Na}{D} .$$

The first term is due to the loss of entropy $Na^2/D^2 + D^2/Na^2$, for $D < a\sqrt{N}$, and the second term is the gain of energy due to the interaction surface-absorbed chain. In good solvent the scaling approach leads to:

$$F/KT = N\left(\frac{a}{b}\right)^{5/3} - \delta\frac{Na}{D} .$$

Minimization gives the de Gennes results:

$$D = a\delta^{-1} \sim \Delta T^{-1} \quad (\theta \cdot s)$$

$$D = a\delta^{-3/2} \sim \Delta T^{-3/2} \quad (g \cdot S) .$$

It is important to note that, in θ solvent, D and l_c (the thickness of the crystalline layer scale) vary in the same way with supercooling. But for different arguments, this fact does not prove that the l_c value is ruled by the thickness of the adsorbed layer; this will be discussed in a forthcoming paper.

The energy of isolated chain adsorbed on the growth face is:

$$F/KT = -1/4N\delta^2 \sim \Delta T^2 \quad (\theta \cdot S)$$

$$= -N\delta^{-5/2} \sim \Delta T^{2.25} \quad (g \cdot S) .$$

In a solution of polydisperse polymer the flux of adsorbed coils is proportional to exp $-F5KT$, and the relation $F/KT = 1$ defines the threshold value:

$$N^* \sim 1/\Delta T^2 \quad (\theta \cdot s)$$

$$N^* \sim 1/\Delta T^{2.25} \quad (g \cdot s) .$$

Chains with $N < N^*$ are very weakly adsorbed and fluctuate easily between adsorbed and bulk situations; these chains cannot crystallize. Chains longer than N^* are absorbed irreversibly on the faces of the monocrystal and crystallize afterwards.

Wunderlich [13, 14] has studied in detail the phenomenon of segregation during crystallization; he found that the critical molecular weight N^* varies as ΔT^{-2}, as is expected from the above relation. We emphasizes that a clear distinction between the exponents 2 (Θs) and 2.25 (gs) is difficult; this is due to the agree of accuracy of the experimental data and the numerous approximations and simplifications of the theory.

Finally, we must note that there is also good agreement between the order of magnitude of the theoretical and experimental values. For a supercooling of ΔT = 10°C, δ = 0.02, $N^* \sim 1000$ for Θ solvents, we compare to the experimental data of Wunderlich on polyethylene $N^* = 1400$ for a similar supercooling. It should be noted that, in this range of molecular weight, the equilibrium melting temperature does not depend on N. In polydisperse polymers containing a fraction of very low molecular weight ($N < 500$) additional effects should be taken in account, in particular the dependence of the melting temperature (and then δT) on the molecular weight [7, 8].

In conclusion, the process of adsorption of polymer chains on growth faces of a monocrystal would explain the segregation effect observed during crystallization. It should be interesting to determine if the kinetics of adsorption, particularly, the kinetics of deformation of the adsorbed chains prior to crystallization have some effect on the kinetics of crystallization.

References

1. Robelin E, Rault J (1989) Macromolecules 22:3581—3594
2. Rault J, Robelin E (1989) J Polym Sci Polym Phys 27:1349—1373
3. Ben Cheick Lorbi F, Hert M, Grenier MF, Rault J (1985) Macromolecules 18:164—170
4. Rault J (1986) CRC Crit Rev Solid Stat Mater Sci 13:57—95
5. Raul J (1990) Ann Chim Fr 15:237-247
6. Hoffman JD, Davis GT (1975) J Research NBS 79a:613
7. Hoffmann JD, Davis JI, Lauritzen JR (1976) In: Treatise on Solid State Chemistry, Plenum, New York 3:497
8. Lauritzen J, Hoffman JD (1960) J Res Natl Bur Stand 64:73—102
9. Yoon D, Flory PJ (1979) Faraday Disc Chem Soc 68:388—390
10. Point JJ, Dosières M (1989) Macromolecules 22:3501—3502
11. de Gennes PG (1969) Rep Prog Phys 32:187—205
12. de Gennes PG (1976) J Phys (Paris) 77:1445—1452
13. Mehta A, Wunderlich B (1974) J Polym Sci Polym Phys Ed 12:255; Mehta A, Wunderlich B (1975) "Colloid and Polym Sci" 253:193
14. Wunderlich B (1979) Faraday Disc Chem Soc 68:239—243

Author's address:

Jacques Rault
Physique des Solides, Bât 510
Université de Paris-Sud
F-91405 Orsay Cedex, France

Polymer nucleating agents: Efficiency scale and impact of physical gelation

A. Thierry, B. Fillon, C. Straupé, B. Lotz, and J. C. Wittmann

Institut Charles Sadron, Strasbourg, France

Abstract: Nucleation of crystallizable polymers is quantified through an efficiency scale. The parameter $E = 100\ ((Tc - Tc_0)/(Tc_{max} - Tc_0))$ where Tc_0 and Tc_{max} are the crystallization temperatures of a non-seeded and of an "ideally" seeded sample, respectively, gives a measure of the efficiency of a given nucleating agent which induces the crystallization of the polymer at Tc. Typical values of E for isotactic polypropylene are given. — The outstanding efficiency of 1,3:2,4-dibenzylidene *D*-glucitol (commonly named 1,3:2,4-dibenzylidene sorbitol (DBS)) as nucleating agent for polyolefins is further analyzed. It is shown that its high efficiency is due to two main features: i) DBS in polyolefins forms thermoreversible gels giving rise on cooling to an extremely fine network of DBS fibers, and ii) the polyolefin crystalline lamellae are able to, subsequently, grow epitaxially on these fibers.

Key words: Nucleating agents; polymer crystallization; polyolefins; 1,3:2,4-dibenzylidene sorbitol; organogels

Introduction

Nucleation of crystallizable polymers by highly divided, interactive foreign substances is a widely used means to enhance the global crystallization rate. It shortens the processing time in injection molding. It reduces the average size of the spherulites and, hence, it improves the mechanical properties and reduces the haze. The nucleating agents have various chemical structures [1, 2]: minerals (talc, etc.), organic salts (e.g., sodium benzoate), organics (e.g., dibenzylidene sorbitol, (DBS)), and they usually act via epitaxial interactions [3].

Generally, the efficiency of nucleating additives is characterized by two different means: either in an isothermal experiment by evaluating the crystallization half-time which is directly related to the nuclei concentration, or in a continuous cooling experiment by measuring the increase in crystallization temperature. Both criteria use a unique reference, the non-seeded state of the polymer.

In this work, we first define a more significant criterion which takes into account not a single reference, but the two extreme reference states attainable: on one side the above "virgin" non-seeded polymer, and on the other side an "ideally" seeded polymer obtainable by self-seeding [4]. Second, we analyze the molecular origin of the known efficiency of dibenzylidene D-glucitol (commonly named dibenzylidene sorbitol (DBS)) to achieve transparent films. The outstanding efficiency of DBS is linked to its dispersion in the polymer. The dispersion is reached through the formation of a thermoreversible DBS gel in the polymer matrix.

Efficiency scale

In a standard differential scanning calorimetry cooling run, a polymer with nucleating additive has a crystallization temperature (Tc) intermediate between that of a "virigin" polymer (Tc_0) and the maximum observable crystallization temperature (Tc_{max}) reached with "ideally" nucleated samples. The latter state can be approached by the self-seeding technique. Via this process, polymer crystal fragments which behave as "ideal" seeds regarding both their chemical and crystallographic interactions, and their state of dispersion can be obtained by controlled partial melting. Self-seeding creates a high density of nuclei (up to 10^{14} fragments/gram) homogeneously dispersed in the polymer matrix. The efficiency (E) of nucleating agents can thus be expressed as a percentage:

$$E = 100\ ((Tc - Tc_0)/(Tc_{max} - Tc_0))\ .$$

The values of E are determined, for example, with heating or cooling rates of 10 K per minute. First, a standard state of crystallization is created through the following process: the sample is completely melted at 483 °K (30 °C above its melting temperature); it is then cooled to 323 °K (50 °C) (at 10 K per minute) where it is kept for 5 min. After subsequent melting or self-seeding of this defined standard state of crystallization, we can measure the different Tc. The standard "virgin" polymer crystallizes subsequently at Tc_0, while under the same conditions the nucleated sample crystallizes at $Tc > Tc_0$. The polymer heated for 5 min at decreasing self-seeding temperatures Ts crystallizes subsequently at increasing temperatures up to a maximum Tc_{max}. The isotactic polypropylene (iPP) sample used in this study was provided by SNEAP/France ($M_w \sim 315 \cdot 10^3$; polydispersity of 5.5). The reference crystallization temperatures Tc_{max} and Tc_0 are respectively 409.1 °K (136.1 °C) and 387.5 °K (114.5 °C).

Table 1. Efficiency coefficient (E) for some known additives of polypropylene:

Additive	C (%)	Tc (°C)	E (%)
4-biphenyl carboxylic acid	2	128.8	66
2-naphtoic acid	1	127.8	62
2,4 dihydroxy, 5-methylpyrimidine (thymine)	1	125.3	50
1,3:2,4-dibenzylidene D-glucitol (DBS)	0.4	123.2	41
4-pyridinecarboxylic acid (isonicotinic acid)	1	123	39
sodium benzoate	1	121.2	31
4-tertiobutyl benzoic acid	1	120.7	29
2,3-pyridine dicarboxylic acid	1	117.3	13

C = additive concentration. Tc = melting temperature of the seeded polypropylene.

In order to evaluate E for the known nucleating agents of iPP, a fine dispersion of the agents in the polymer matrix is prepared through dissolution of the polymer and the additives in a common solvent (xylene). Typical values of E are reported in Table 1. They range from 30% to 60% (e.g., DBS: 40%); values as high as 70% have been observed [5].

Compared to the two methods already mentioned, the proposed efficiency scale is a substantial improvement for a valid classification of potential nucleating agents. However, it cannot take into account all the phenomena involved in the polymer solidification during processing, in particular, the flow-controlled phenomena. As, however, the solidification process during injection molding is nucleation-rate-controlled [6], a better selection of nucleating candidates is useful.

Impact of physical gelation

DBS is known to be a first-rate nucleating agent, producing nearly transparent polyolefin films. It is also known to be a very efficient gelifying agent, even at very low concentrations, and for a large variety of polar or nonpolar media [7].

What is behind these properties? What can be learned from them?

DBS, a butterfly-shaped molecule (Fig. 1), is chiral and amphiphilic in character; through hydrogen bonding of the nonreacted OH groups it is able to make self-assembling superstructures. The wings of the buttefly, i.e., the hydrophobic aromatic rings, ensure a good compatibility of DBS and polyolefins at high temperatures. Thus, in the

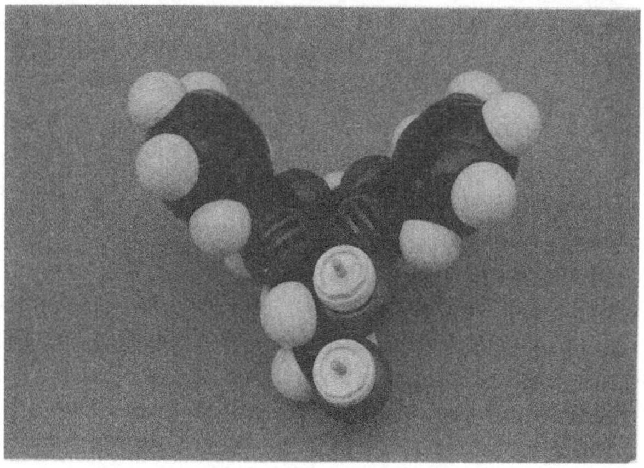

Fig. 1. DBS molecular model showing the buttefly shape with the apolar aromatic wings and the polar sugar body

molten state, DBS molecules are completely dissolved and homogeneously dispersed in the polymer matrix. When this homogeneous mixture is cooled, a transient gel is formed. Like many organogels already studied [8–9] its building elements are nearly endless, thin, twisted fibers, sometimes cooperatively twisted together (Fig. 2). The diameter of the elementary fiber is typically 10 nm and, hence, its specific area is close to 400 m^2/g.

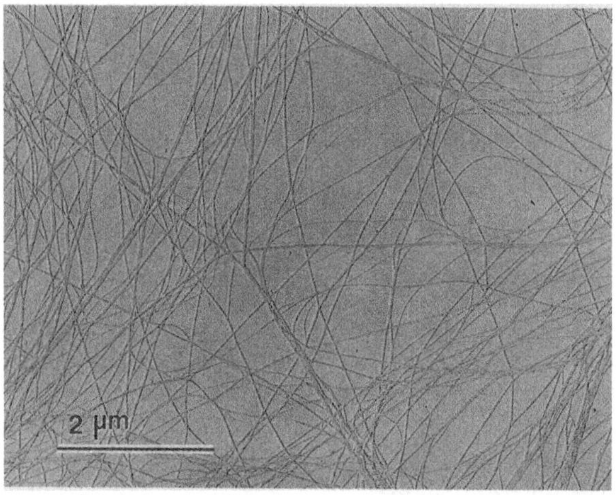

Fig. 2. Electron micrograph of DBS twisted fibers; the fibers formed directly on a carbon-coated grid from a THF/benzene mixture are Pt/C shadowed

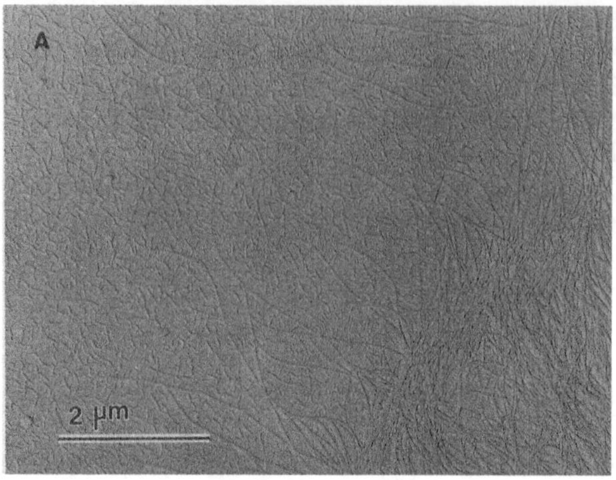

Fig. 3A: Electron micrograph of polyethylene-decorated DBS fibers deposited as in Fig. 2 (Pt/C shadowed)

Fig. 3B. Electron diffraction pattern of polyethylene vapor-deposited on DBS (002 reflection arrowed)

Specific interactions of DBS with polyolefins enable the formation of a thermoreversible gel in the polymer matrix. However, this feature alone is insufficient to explain the outstanding nucleating performance of DBS. As already mentioned, Wittmann and Lotz [3] demonstrated that epitaxial interactions favor a high degree of nucleating efficiency. This last requirement is apparently also fulfilled in the case of the DBS-polyolefins pair. Figure 3A shows regularly twisted DBS fibers decorated with polyethylene, following the established polymer-decoration method [10]; the polyethylene lamellae oriented on and by DBS fibres are clearly visible. This first indication of epitaxy is supported by the diffraction pattern of Fig. 3B, which shows a unique orientation of the *c* axis of polyethylene vapor-deposited on a bundle of DBS fibers.

Thus, DBS has all the characteristics required for an efficient nucleating agent. The thermoreversible gel formed on cooling creates a dense network of crystallized fibers, i.e., of nucleating sites on which polyolefins can epitaxially nucleate.

Hence, in addition to *epitaxial interactions*, *physical gelation* which creates an outstanding state of dispersion of the additive in the polymer matrix should be sought in developing new classes of nucleating agents.

References

1. Beck HN (1967) J Appl Polym Sci 11:673
2. Binsbergen FL (1977) J Polym Sci Polym Symp 59:11

3. Wittmann JC, Lotz B (1981) J Polym Sci, Polym Phys Ed 19:1837
4. Blundell DJ, Keller A, Kovacs A (1966) Polymer Letters 4:481
5. French patent 90:00046
6. Eder G, Janeschitz-Kriegl H, Liedauer S (1990) Prog Polym Sci 15:629
7. Japan patent 85:16909 and 86:68405
8. Tachibana T, Mori T, Hori K (1981) Bull Chem Soc Japan 54:73
9. Wade RH, Terech P, Hewat EA, Ramasseul R, Volino FJ (1986) Colloid Interface Sci 114:442
10. Wittmann JC, Lotz B (1985) J Polym Sci, Polym Phys Ed 23:205

Authors' address:

A. Thierry
Institut Charles Sadron
6, rue Boussingault
67083 Strasbourg, Cedex, France

Regime III crystallization in poly-(l-lactic) acid

S. Mazzullo, G. Paganetto, and A. Celli

Himont Italia, "G. Natta" Research Centre, Ferrara, Italy

Abstract: The theory developed by Hoffmann for regime III crystallization is applied to poly-(l-lactic) acid bulkd growth rates of crystallization from the molten and the glassy states. — As the temperature decreases a marked upward change in the slope of the growth rate versus temperature curves is observed, which is interpreted as a regime II to regime III transition. The regime transition occurs around 140 °C which corresponds to an undercooling of 67 °C.

Key words: Regime crystallization; poly-(l-lactic) acid; kinetic theory of nucleation; biodegradable polymers

Introduction

Polymers of lactic acid are obtained by catalytic ring opening polymerization of di-lactides of lactic acid. Di-lactides exist in three stereo-isomeric forms: dextro, levo, and dexro-levo forms. This paper examines polymers resulting from polymerization of the levo form (PLLA).

PLLA is a semicrystalline polymer which is bioresorbable when implanted into a living organism. The main objective of this investigation was to determine the nucleation constants for the bulk growth crystallization of PLLA from the melt and from the glass.

Materials and method

A commercially available PLLA was used (RESOMER L214, Boehringer, Ingelheim, FRG), having a viscosimetric molecular weight (Mv) of 700 000 g/mol. The polymer was processed as obtained from the manufacturer, without any additional purification.

Bulk crystallization growth experiments were conducted on thin flakes of polymer sandwiched between two glass microscope slides placed on a Zeiss optical microscope equipped with a Mettler hot stage (FP52). Bulk growth rates were detected by light depolarization measurements, using a photomultiplier inserted in th eyepiece tube of the microscope; they were recorded by a writing chart recorder [1].

In the case of crystallization from the melt, the samples were heated to 10—15 °C above the equilibrium melting point (Tm° = 480 K [2]), kept at that temperature for 10 min, and then cooled rapidly to the required isothermal crystallization temperature.

In the case of crystallization from the glassy state, the first step was to prepare amorphous glassy flakes. This was easily done by quenching of the molten polymer film in liquid nitrogen. The second step was to rapidly heat the glassy flake to the required isothermal crystallization temperature. Final morphology after crystallization from both the glass and the melt is shown in Fig. 1.

This experimental procedure closely follows the one described in the literature [2] for the same polymer.

The only difference is in the measured parameter, which in this paper refers to the bulk growth rate of crystallization as a function of temperature, while in [2] it is referred to the radial growth rate of spherulites.

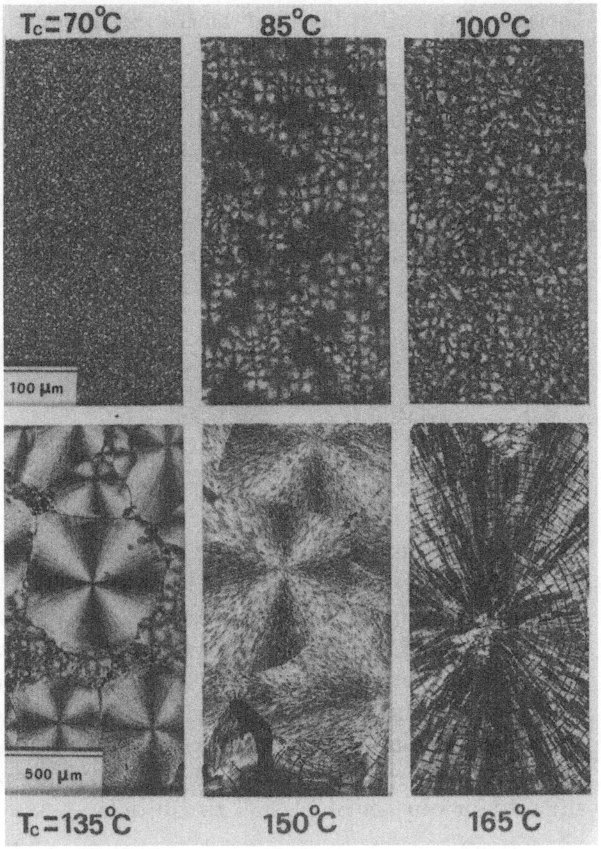

Fig. 1. Spherulitic morphology of PLLA crystallized from the glass (Tc = 70°, 85°, 100°C) and from the melt (Tc = 135°, 150°, 165°C)

Results and discussion

According to Hoffmann's formulation of kinetic theory of crystallization [3], the bulk growth rate of a linear polymer crystal with chain folds is described within a given regime by the expression:

$$k(T) = k_0 \exp - (u^*/R)/(T - T_\infty)$$

$$\times \exp - Kg/(T(Tm^\circ - T)f) .$$

Where $k(T) = 1/t_{1/2}$ and $t_{1/2}$ is the time required to attain 50% of crystallization; k_0 is the preexponential factor containing quantities not highly dependent on temperature; u^* is a "universal" constant characteristic of the activation energy of chain motion in the crystallizing medium. R is the gas constant; $T_\infty = Tg - 30$ is the theoretical temperature at which all motion associated with viscous flow ceases, where Tg is the glass transition temperature; Tm° is the equilibrium melting temperature and $f = 2T/(Tm^\circ + T)$ is a correction factor that accounts for the change of the heat of fusion.

Our experimental half-times of crystallization are compiled in Table 1 (crystallization from the melt) and in Table 2 (crystallization from the glass). The use of bulk growth instead of radial growth data in order to find regime transitions is supported by several authors [4]. The use of the same plot of crystallization data from the glass or from the melt is justified by the following two considerations: 1) the viscosity of the medium is closely related to the density of crystallization nuclei; 2) wide-angle x-ray diffraction does not show crystallinity in the glassy flakes before crystallization. Therefore, we assumed that the transport term is described by the universal constant u^*, independent of the path followed to reach the crystallization temperature. Three regimes of polymer crystallization have been proposed [3] on both theoretical and experimental grounds: regime I at low undercooling ($Tm^\circ - T$), regime II at average, and regime III at high undercooling.

Table 1. PLLA crystallized from the melt: Half-time of crystallization as a function of temperature

T (°C)	$t^{1/2}$ (s)	T (°C)	$t^{1/2}$ (s)
110	228	145	9828
115	249	150	10800
120	372	150	19800
125	576	155	9720
130	1512	155	23940
135	2412	160	36360
135	3140	160	52020
135	2988	165	68040
140	4284	—	—

Table 2. PLLA crystallized from the glass: Half-time of crystallization as a function of temperature

T (°C)	$t^{1/2}$ (s)	T (°C)	$t^{1/2}$ (s)
100	150	80	7884
95	300	75	25920
90	678	70	165600
85	2304		

Vasanthakumari and Pennings [2], by means of radial growth crystallization data, were able to find regime I and regime II (transition temperature, T_b = 163 °C) for a PLLA having a *Mv* of 150 000 g/mol.

In this work, by means of bulk growth crystallization experiments, we were able to find regime II again and, for the first time, the regime III crystallization for a PLLA having *Mv* = 700 000 g/mol.

The transition temperature from regime II to regime III occurs at 140 °C, as shown by the intersection point of the two straight lines in the Hoffman identification plot of Fig. 2. At this temperature, the previous work of Vasanthakumari and Pennings [2] shows that crystallization is still in regime II. However, there is no discrepancy between our results and those of the above authors, because their data did not go high enough into the temperature range of regime III to clearly define the transition.

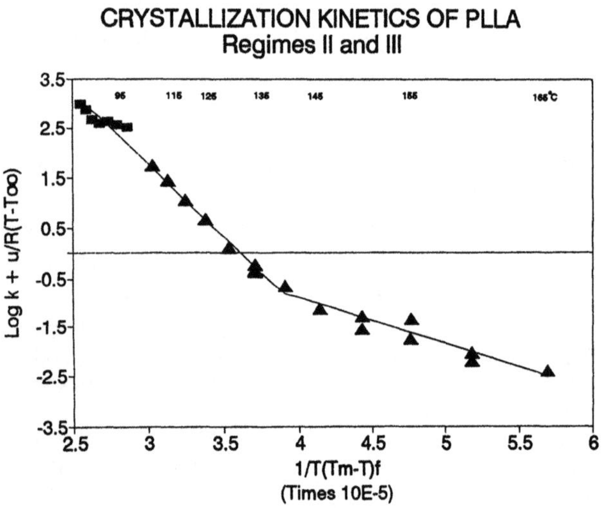

Fig. 2. Hoffman identification plot showing regime II and regime III in PLLA

Parameters and identified kinetic constants are compiled in Table 3. The nucleation constant *Kg*, in regime II, is in agreement with the literature [2], while the nucleation constant in regime III is some 50% higher than was predicted by theory. This discrepancy may be caused by at least three factors which heavily influence the numerical analysis of experimental data: these factors are the actual values of the parameters u^*, $T\infty$, and $Tm°$. For the sake of comparison with the literature, we used the same set of values of Vasanthakumari and Pennings [2]. However, this choice does not seem suitable for our set of crystallization data.

Table 3. Thermodynamic and kinetic parameters for PLLA

Mv 10^{-5}	u^* (J/mol)	*Tg* (K)	Remark
7.0	6280[a])	333[a])	Regime II
Tm° (k)	k_0 (s^{-1})	*Kg* 10^{-5} (K^2)	Remark
480[a])	686	2.16	Regime II
	3.87 10^{10}	6.76	Regime III

[a]) Reference [1].

References

1. Binsbergen FL, De Lange BGM (1970) Polymer 11:309–332
2. Vasanthakumari R, Pennings AJ (1983) Polymer 24:175–178
3. Clark EJ, Hoffman JD (1984) Macromolecules 17:878–885
4. Hoffman JD, Miller RL (1989) Macromolecules 22:3502–3505

Authors' address:
Dr. S. Mazzullo
Himont Italia
"G. Natta" Research Centre
44100 Ferrara, Italy

Progress in Colloid & Polymer Science Progr Colloid Polym Sci 87:35—38 (1992)

Comparative study of isomeric polyamides with isolated methylene groups: Nylons 1,6 4,3 and 2/5

J. Puiggalí, J. E. Aceituno, L. Franco, J. Lloveras, A. Prieto, X. Vidal, A. Xenopoulos, J. M. Fernández-Santín, and J. A. Subirana

Departament d'Enginyeria Química, ETSEIB, Universitat Politècnica de Catalunya, Barcelona, Spain

Abstract: Three isomeric polyamides have been studied by calorimetry, electron and x-ray diffraction. They only differ by the relative orientation of their amide groups. Each of them has a unique structure, different from conventional polyamides. Nylon 2/5 is similar to polyglycine II, nylon 1,6 has a modified γ structure and nylon 4,3 may have an intermediate structure.

Key words: Polyamides; nylon 1,6; nylon 4,3; nylon 2/5; x-ray diffraction; electron diffraction

Introduction

We have studied several aliphatic polyamides [1—6] which contain isolated methylene groups placed between two amide groups. In many cases these polymers acquire unique helical conformations which are not usually found in polyamides. They can be grouped into two families, either as derivatives of 1,n nylons, or as sequential copolymers of glycine and ω-amino acids (2/n nylons). Paiaro et al. [7] have also described a related family of polyamides, the family of n,3 nylons, which can be considered to be derivatives of malonamide. Thus, there are three families of polyamides with isolated methylene groups, and we decided to compare the properties of three isomers, one of each series:

Nylon 1,6 $(CONHCH_2NHCOCH_2CH_2CH_2CH_2)_n$;

Nylon 4,3 $(NHCOCH_2CONHCH_2CH_2CH_2CH_2)_n$;

Nylon 2/5 $(CONHCH_2CONHCH_2CH_2CH_2CH_2)_n$.

The method of synthesis is quite different in each case. In summary, nylon 1,6 is obtained by the reaction of formaldehyde with a dinitrile [8], nylon 4,3 by the reaction of carbon suboxide with a diamine [7] and nylon 2/5 by conventional methods of peptide chemistry with active esters as starting materials [4].

Characterization of polymers

Elementary analysis, as well as infrared and NMR spectra agree with the molecular formulae given above, without any signs of chain branching. The molecular weights were estimated from the intrinsic viscosity, using the equations reported by Elias and Schumacher [9] for nylon 66. The intrinsic viscosities were measured in two different solvents: dichloroacetic acid, which is a bad solvent for nylon 66 and induces aggregation, and 90% formic acid, which is a good solvent. The results obtained are given in Table 1. It is obvious that the estimated molecular weights depend on the solvent used, which indicates that the hydrodynamic behavior of these polymers differs from nylon 66. In any case, it is clear that nylon 4,3 has a molecular weight larger than the other two polymers.

Calorimetry

The melting curves obtained from fragments of vacuum dried crystalline mats in a Perkin Elmer DSC 4 instrument in a nitrogen atmosphere at

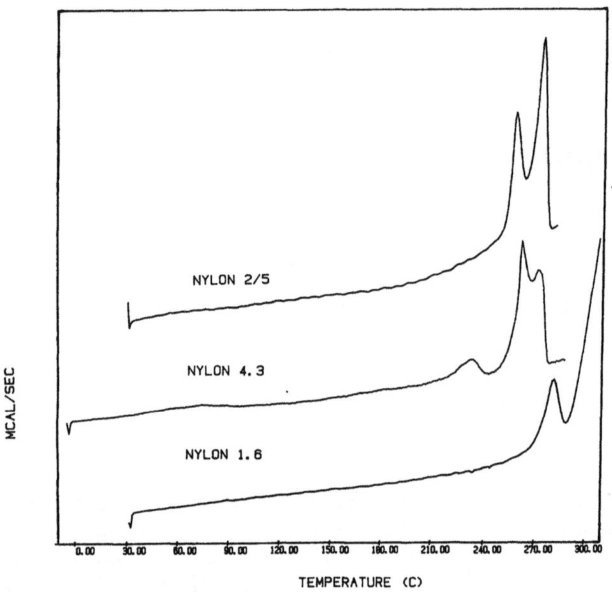

Fig. 1. Thermograms obtained from crystalline mats of the three compounds studied in this paper. Nylon 1,6 decomposes before melting, as discussed in the next

Table 1. Molecular weights estimated from intrinsic viscosities. The solvents used were dichloroacetic acid (DCA) and 90% formic acid (FOR). The results are discussed in the text

Nylon	DCA	FOR
4,3	6000	5300
1,6	2700	1300
2/5	800	2200

10°/min are shown in Fig. 1. Nylons 2/5 and 4,3 show multiple melting peaks, as is often found in other polyamides. The highest melting point, which characterizes the best crystals, is 275°C for both polymers. On the other hand, nylon 1,6 decomposes before melting. No melting point could be measured, but its value should be higher than 300°C. The higher melting point of the latter compound suggests a different molecular structure, in agreement with the results that we will present below. Its thermal decomposition might be related to the eventual tendency of this compound to evolve formaldehyde upon heating, as in the reaction opposite to its synthesis.

The melting enthalpy is 15.7 kJ/mol for nylon 4,3 and 19.9 kJ/mol for nylon 2/5. Using the additivity concept of van Krevelen (10), we can estimate a crystallinity of about 63% and 80% for both polymers, respectively.

Electron microscopy

Single crystals were obtained by isothermal crystallization in the range 90°—130°, in either dilute 1,4 butanodiol solution (nylons 1,6 and 4,3) or in formic acid/butanol mixture (nylon 2/5). The crystals were small and elongated for nylon 1,6, triangular for nylon 2/5, and with a lozenge shape for nylon 4,3. The latter crystals very often aggregated in an hexagonal fashion. The electron diffraction patterns obtained from such crystals are shown in Fig. 2. The patterns of nylons 2/5 and 4,3 appear to have hexagonal symmetry, whereas the pattern of 1,6 does not have this symmetry, since the second-order reflection only appears in one direction. In all cases the spacing of the first order reflection is about 0.415 nm.

X-ray diffraction

Diffraction patterns were obtained from mats covered with calcite as an internal standard and are shown in Fig. 3. In the case of nylon 2/5, we could not obtain an oriented mat: the unoriented powder diagram is shown in Fig. 3. All patterns have similar equatorial and chain spacings, but the lamellar thickness varies in each case, as shown in Table 2. The pattern of nylon 1,6 has two features which distinguish it from the other two materials:

— The meridional chain-spacing intensity is sampled by the lamellar thickness, which indicates a rather regular thickness of the lamellar crystals. Thus, in the medium angle region three meridionally oriented rings are observed at spacings 2.38, 1.17, and 0.95 nm, which correspond to the second, fourth, and fifth order of the 4.7 nm lamellar spacing. In this case the chain spacings has been determined from powder samples, which show a single ring with a 1.01 nm spacing.

— The equatorial ring at 0.418 nm spacing is accompanied by an additional ring at about 0.403 nm, which confirms that this pattern is not hexagonal, in agreement with the electron diffraction results shown in Fig. 2.

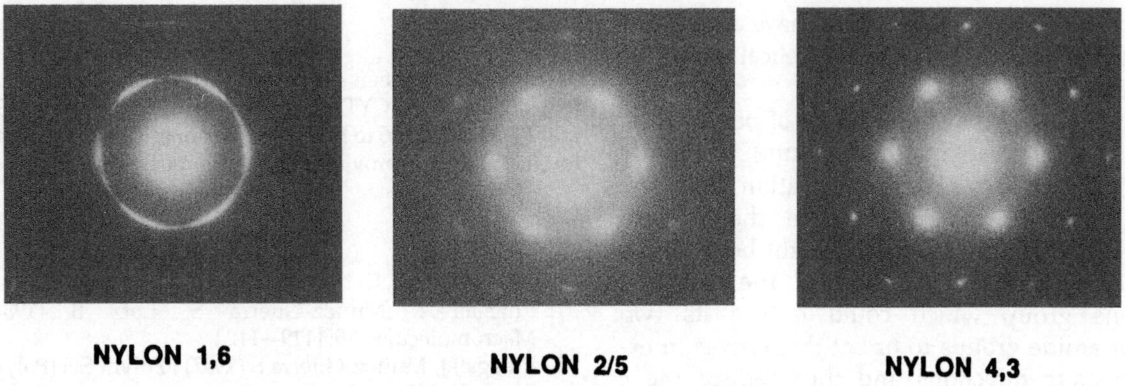

Fig. 2. Electron diffraction patterns obtained from a single crystal of the three polyamides. In the case of nylon 1,6 a group of approximately parallel crystals was used

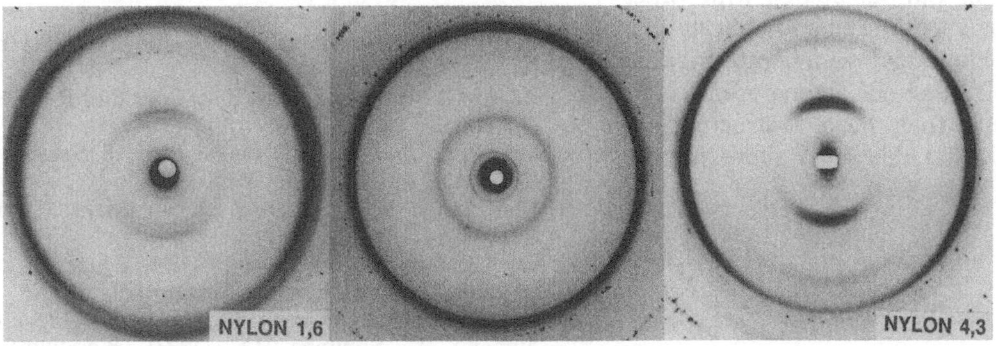

Fig. 3. X-ray diffraction patterns obtained from a crystalline powder of nylon 2/5 and from partially oriented crystalline mats of the other two polyamides. The strongest ring corresponds to a spacing of about 0.415 nm in the three cases. The spotty rings in nylons 2/5 and 4,3 are due to the calcite used as a reference

Table 2. Summary of x-ray spacings (nm)

Nylon	Equatorial	Chain	Lamellar
4,3	0.413	1.00	6.3
2/5	0.415	1.01	4.8
1,6	0.418/0.403	1.01	4.7

— The meridional chain-spacing intensity is sampled by the lamellar thickness, which indicates a rather regular thickness of the lamellar crystals. Thus, in the medium angle region three meridionally oriented rings are observed at spacings 2.38, 1.17, and 0.95 nm, which correspond to the second, fourth, and fifth order of the 4.7 nm lamellar spacing. In this case the chain spacings has been determined from powder samples, which show a single ring with a 1.01 nm spacing.

— The equatorial ring at 0.418 nm spacing is accompanied by an additional ring at about 0.403 nm, which confirms that this pattern is not hexagonal, in agreement with the electron diffraction results shown in Fig. 2.

Structural interpretation

Nylon 2/5 apparently has a truly hexagonal (or trigonal) symmetry and its structure is similar to that found in other copolymers of glycine (1,4), which have been shown (11,12) to have a structure very similar to the threefold helical form of polyglycine II.

Nylon 1,6 belongs to the γ family of polyamides, but its chain repeat length (1.01 nm) is slightly shorter than expected by extrapolation from the values found in other nylons in the γ form (1.03—1.04 nm). This difference might be due to a particular conformation around the isolated methylene group which could induce its two neighbor amide groups to orient themselves in exactly opposite directions and thus reduce the c spacing from the usual γ value.

Nylon 4,3 has an apparent hexagonal symmetry in its electron diffraction pattern (Fig. 2), but the crystal morphology does not show such symmetry. Work in progress in our laboratory with other polyamides of the n,3 series seems to indicate that these polymers may have hydrogen bonds oriented in two directions placed at 60° among themselves. These nylons would then have a structure intermediate between that described above for the isomeric polymers 1,6 and 2/5.

Discussion

The three isomeric polyamides described in this paper each have a unique structure which differs from the conventional α and γ structures of most polyamides. Their conformation is intermediate between the γ form and the threefold helix of polyglycine II. It appears that an isolated methylene group placed between two amide residues has strong conformational preferences which depend on the relative disposition of the two amide groups, a question which will be analyzed in more detail elsewhere (13).

Acknowledgements

This work has been supported by the Brite-Euram Project BE-3106-89, CICYT grant MAT89-1136-CE and by NSF grant DMR 8816326 to Prof. M. Goodman. We are grateful to Dr. Paiaro for providing a sample of nylon 4,3.

References

1. Puiggalí J, Muñoz-Guerra S, Lotz B (1986) Macromolecules 19:1119—1124
2. Puiggalí J, Muñoz-Guerra S (1987) J Polym Sci (Polym Phys) 25:513—523
3. Puiggalí J, Muñoz-Guerra S (1987) J Polym Sci (Polym Chem) 25:1445—1448
4. Puiggalí J, Muñoz-Guerra S, Subirana JA (1987) Polymer 28:209—212
5. Muñoz-Guerra S, Fita I, Aymamí J, Puiggalí J (1988) Macromolecules 21:3364-3468
6. Puiggalí J, Muñoz-Guerra S (1989) J Polym Sci (Polym Phys) 27:1563—1567
7. Paiaro G, Pandolfo L, Busico V, Corradini P (1988) Eur Polym J 24:99
8. Magat EE, Chandler LB, Faris BF, Reith JE, Salisbury LF (1951) J Am Chem Soc 73:1031
9. Elias H-G, Schumacher R (1964) Makr Chem 76:23—52
10. Van Krevelen DW, Hoftyzer PJ (1976) Properties of Polymers, 2nd ed., Elsevier, New York
11. Tormo J, Puiggalí J, Vives J, Fita I, Lloveras J, Bella J, Aymamí J, Subirana JA (1992) Biopolymers 32:643—648
12. Bella J, Puiggalí J, Subirana JA (1991) In: Giralt E, Andreu D (eds) Peptides 90. ESCOM Science Publishers, Leiden, pp 452—453
13. Bella J, Subirana JA (1992) Polymer Preprints 33(1):276—277

Authors' address:

Prof. Juan A. Subirana
Departament d'Enginyeria Química
ETSEIB
Av. Diagonal 647
E-08028 Barcelona, Spain

Thermodynamic anomalies in the pressure-temperature phase diagram of poly-4-methyl pentene-1: Disordering of crystals on cooling

S. Rastogi, M. Newman, and A. Keller

H. H. Wills Physics Lab., University of Bristol, U.K.

Abstract: We report on some highly unusual phase behavior of general implication for condensed matter in a polymer poly-4-methyl pentene-1. In situ x-ray studies, at the prerequisite pressure and temperature, reveal amorphization in this polymer, passing through a continuously varying sequence of mesomorphic states, at two widely separated temperature regions, the process being reversible. Present findings invoke the possibility of the reentrant, widely separated disordered phases with temperature. Such possibility has arisen for the first time in a single-component system. The pressure-induced disordered phase, obtained at room temperature, converts into crystal with increasing temperature and reverts to a glassy disordered phase on lowering the temperature. Other unexpected findings include a new pressure-induced modification of poly-4-methyl pentene-1, with a one-way entry only. A few aspects of similar features found in other materials are also stated.

Key words: Poly-4-methyl pentene-1; phase diagram; pressure; temperature; amorphization

The purpose of this communication is to report some highly unusual observations in a polymer poly-4-methyl pentene-1 (P4MP1), specifically, phase transitions induced by pressure (*P*) and temperature (*T*). Polymer crystallizes in tetragonal phase (C_t) on straightforward solidification from the melt. At room temperature and atmospheric pressure crystal density (P_c = 813 kg m^{-3}) is less than the amorphous density (P_A = 830 kg m^{-3}), whereas above 50 °C, at atmospheric *P*, such an inverted relationship, at room temperature, acquires special significance, as discussed in the following.

Figure 1, shows effect of pressure at room temperature on an oriented film of P4MP1. In situ x-ray observations reveal that on raising *P* above a threshold value, crystal reflexions broaden and become diffuse, those close together merge into liquid-pattern-type blobs, while others disappear. Indeed, long-range crystal periodicity is lost, and the system becomes disordered (Figs. 1a—d). The whole sequence is fully reversible with *P* (Fig. 1e). The continuous effect of pressure-induced amorphization (denoted as LC, A) is shown as Pathway 1 in Fig. 2. Figs. 1a—d refer to points (i)—(iv) in Fig. 2. At the highest *P*(~6Kb) accessible to us, point (v) in Fig. 2, there is virtually an oriented amorphous material.

The above change from order to disorder with increasing *P* is consistent with the anomalous density relation $P_c < P_A$. In this respect, P4MP1 seems to fall in the same class as water or bismuth where *P* causes a decrease in the melting point. However, the presently observed transitions occur about 220 °C below the a.p. melting point and not in the melting region as in the familiar case of water.

To bridge this latter gap we proceeded to examine the effect of *P* also in the range of melting temperature (T_m), i.e. around 245—300 °C, along Pathway 2 in Fig. 2.

Experimental findings shown elsewhere [1, 2] clearly demonstrate that on raising the *P*, isothermally *T* (248 °C), from point (vi) to (vii) in Fig. 2, initially C_t goes into the liquid phase (referred as *LC′*, *A′*). This process is completely reversible. This indicates a decrease in the melting temperature (T_m) with *P*. Experiments [1, 2] conclusively

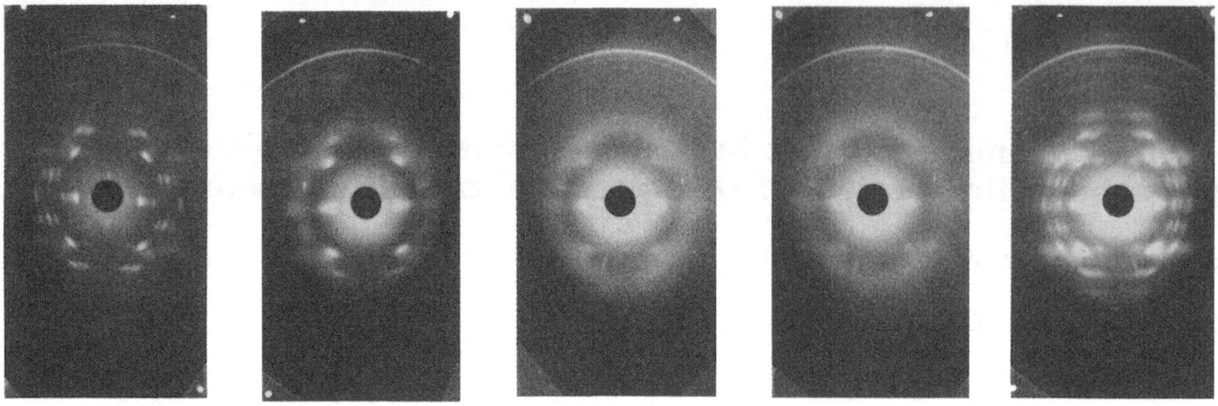

a b c d e

Fig. 1. X-ray diffraction patterns of drawn films of P4MP1 at r.T. recorded in situ at the various pressures: a) 0.24 Kb; b) 2.24 Kb; c) 4.56 Kb; d) 5.36 Kb; e) 0.24 Kb. (a) to (d) correspond to points (i) to (iv) respectively in Fig. 2 with (e) corresponding to a reversal to point (i). The sharp reflexions on the periphery are due to diamond and the rings to a gasket; these serve as useful intensity standards when comparing different members in a sequence

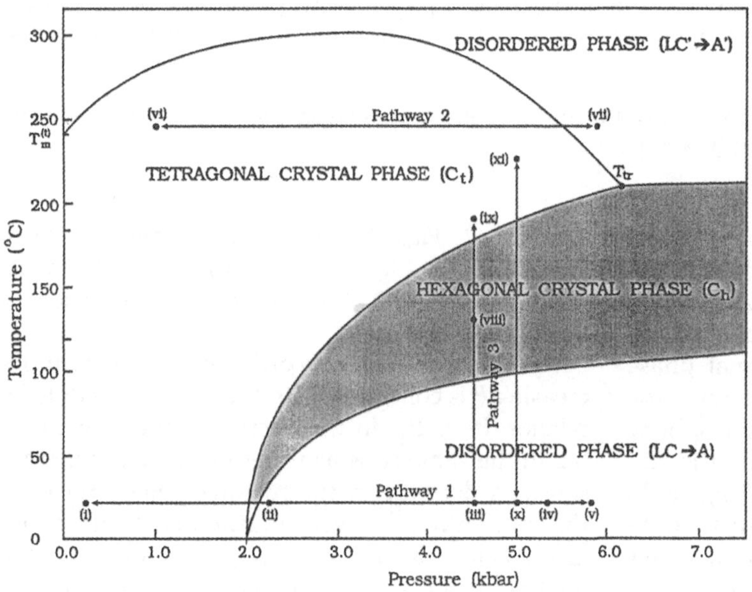

Fig. 2. *P-T* phase diagram of P4MP1. Liquid (amorphous), solid (crystal in tetragonal form) and intermediate (liquid crystal) denoted by *A*, C_t and *LC* respectively. The shaded region encompasses a further (hexagonal) crystal phase

demonstrate that T_m first rises with *P*, reaching a maximum at a certain P^1, decreasing thereafter on further increase in *P*. This is readily explained on the basis that the amorphous (liquid) phase is more compressible than the crystal. This in turn means that, when starting from $\rho_c > \rho_A$, the densities will invert at a sufficiently high *P*, leading to the condition of $\rho_c < \rho_A$.

Two widely separated disordered phased in the different temperature regions on *T* axis imply a liquid phase reentry — the first ever anticipated in a single-component system. Such a possibility of reentry is indeed verified along Pathway 3 in Fig. 2.

Experiments [1, 2] along Pathway 3 show that the disordered phase achieved along Pathway 1, (i) to (x), crystallizes in C_t, on heating from (x) to (xi) for the fixed *P*. Crystalline material C_t goes back into the disordered phase on lowering the *T* from (xi) to (x), suggesting that C_t has higher entropy compared to that of the disordered phase (*LC, A*). To our knowledge, such a process of disordering on cooling has no precedent in any known system.

The shaded region in Fig. 2 corresponds to a new crystalline hexagonal (C_{th}) phase. It may be obtained along Pathway 3, i.e. on raising *T* for the fixed *P* from (iii) to (viii). On raising *T* further, from (viii) to (ix), the C_h to C_t transition occurs. Such a solid-to-solid transition is a one-way entry on the *T* axis in the *P-T* phase diagram, i.e. on cooling from (ix) to (viii) to (iii), C_t goes directly into the disordered phase (for details see [2]).

The above findings are novel and are the first ever reported for a polymer. Some aspects of this behaviour (like *P*-induced amorphization with an inversion in sign of the temperature dependence of T_m with *P*) have been found or surmised recently in water-ice [3] and silica [4].

Amorphization with *P* at room temperature is reported in several materials, such as SiO_2 (quartz) [5], $AlPO_4$ [6—8], SnI_4 [9], $LiKSO_4$ [8], $Ca(OH)_2$ [10—13], $CaAl_2SI_2O_8$ (anorthite) [9] and Fe_2SiO_4 (fayalite) [9]. In most of these materials pressure-induced amorphization is irreversible, but in some cases reversibility could also be observed, such as in SnI_4, $AlPO_4$, $LiKSO_4$, $Ca(OH)_2$.

References

1. Rastogi S, Newman M, Keller A (1991) Nature 353:55—88
2. Rastogi S, Newman M, Keller A (1991) J Polym Sci Phys Ed (In press)
3. Mishima O, Calvert LD, Whalley E (1984) Nature 310:393—395; ibid (1985) 314:76—78. Hemley RJ, Chen LC, Mao HK (1989) Nature 338:638—640
4. Hemley RJ, Jephcoat AP, Mao HK, Ming LC, Manghnani MH (1988) Nature 334:52—54
5. Hemley RJ, Jephcoat AP, Mao HK, Ming LC, Manghnani MH (1988) Nature 334:52—54
6. Jayaraman A, Wood DL, Maines RG (1987) Phys Rev B35:8316—8321
7. Williams Q, Jeanloz R (1988) Science 239:902—905
8. Kruger MB, Jeanloz R (1990) Science 249:647—649
9. Williams Q, Jeanioz R (1989) Nature 338:413—415
10. Fujii Y, Kowaka M, Onodera A (1985) J Phys C18:789—797
11. Sugai S (1985) J Phys C18:799—808
12. Meio FEA, Lemos VE, Cerdeira F, Mendes F (1987) J Phys Rev B35:3633—3636
13. Sarkaran H, Sikka SK, Sharma SM, Chidambaram R (1988) J Phys Rev B38:170—173

Authors' address:

A. Keller
H. H. Wills Physics Lab.
University of Bristol
Bristol BS8 1TL, U.K.

The role of transient hexagonal phase in crystallization of polymers

S. Rastogi, M. Hikosaka, A. Keller, and G. Ungar

H. H. Wills Physics Lab, University of Bristol

Abstract: Crystallization studies of polyethylene (PE) under hydrostatic pressure substantiate the earlier findings by Bassett and Turner [1, 2] that crystallization occurs initially in the hexagonal (h) phase, even within the thermodynamically stable orthorhombic (o) region. Present studies demonstrate that this is, in fact, a necessary condition for crystallization, at least in the temperature and pressure range in question; in fact, once a crystal transforms from h to o phase, further growth of crystal stops. Morphological features show that crystal growth includes both lateral and thickening growth; in the latter case the chains thicken continuously from folded to extended form and beyond. It is the thickness in a crystal which governs the phase inversion from h to o phase. Thickness dependence on phase inversion is explicitly shown in the polymer trans 1,4-polybutadience, where the mobile h phase, similar to PE, can exist, even at atmospheric pressure.

Key words: Polyethylene; trans 1,4-polybutadiene; transient hexagonal; phase crystallization; mesophase

Crystallization of polyethylene (PE) at elevated pressures, above and below the triple point in the established P-T phase diagram [1, 2] leads to the formation of extended chain crystals [3], a consequence of the chains being in the hexagonal (h) phase. This phase is thermodynamically stable within Region I of the P-T phase diagram (Fig. 1). Volumetric and thermal measurements revealed a transient phase within the o stability regime, attributed to the h phase, which then can transform into the final stable phase (o) subsequently [2]. In subsequent works changes in the optical appearance of the crystals were attributed to such an h → o transition [4] as observed on cooling [1].

In the present work, using a piston-cylinder type diamond anvil cell (operating both at constant temperatures *and* pressures, even during a phase transformation) and a sharp fraction of polyethylene (M_w = 32 000 supplied by NBS), we followed crystallization in situ, optically, under isothermal *and* isobaric conditions. As in preceding works the crystals were seen as birefringent entities between crossed polaroids forming and growing with time. Using the optical criterion [1, 4] for the hexagonal (h) and orthorhombic (o) crystal structures (uniformly smooth and irregular blodgy bright streaks, respectively), we could follow the crystallization in preselected regions of the phase diagram in terms of size of crystal and nature of crystal structure. In the course of this the dominant role of the h phase became increasingly apparent. Not only does it occur during crystallization, even within the o stability regime, but is a state necessary for the crystallization to start and proceed, at least within the P- and T-range examined here, even when, in the o regime, it corresponds to a metastable state.

The above can be clearly seen from Fig. 2 which demonstrates crystal growth within the o regime "below" the triple point. Here, smooth birefringent entities are seen to form and grow with time. We know from separate electron microscope work [5, 6] that these entities are isolated single-crystal layers. After a certain crystallization time (Fig. 2c) the existence of two distinct types of crystal become apparent, like crystals A and C, having a well-defined

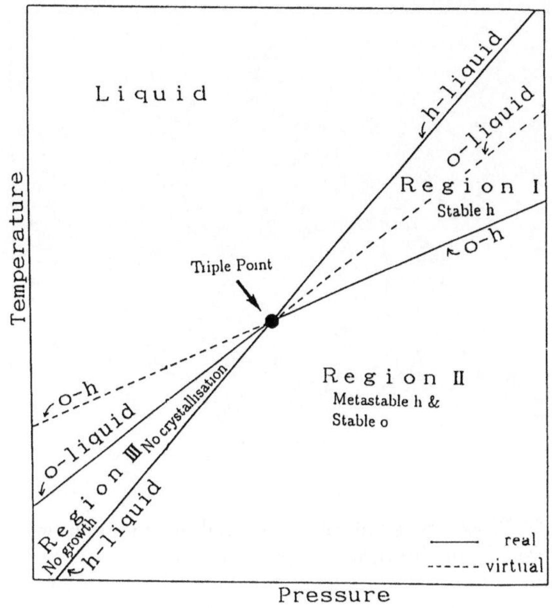

Fig. 1. Schematic P-T phase diagram including real and virtual boundaries

boundary, and crystals B and D, having distorted shape. That such crystals are indeed intrinsically different becomes evident on raising the temperature at fixed pressure (Fig. 2d). Here, crystals having well-defined boundary disappear, crystals A and C, leaving behind distorted crystals such as B and D, which means that the former have a lower melting point. This alone provides self-contained evidence that the smooth crystals are metastable with respect to those of irregular appearance in that particular regime of the phase diagram. In situ x-ray studies [5] then revealed further that the crystals having well defined boundaries are indeed in the h form, whereas those with distorted boundaries are in the o phase, consistent with the preceding optical attribution (Fig. 2).

From the above criterion of optical distinction between h and o crystals the following conclusions may be drawn from evidence such as in Fig. 2.

a) Fig. 2a shows that within the P and ΔT range concerned, even within the thermodynamically stable o phase, crystals only form and grow in the h phase initially, even within the thermodynamically stable state.
b) When in the o stability regime, after certain crystallization time, a crystal initially in h phase transforms to the o phase, like crystals B and D.
c) Once a crystal transforms from h to o phase it stops growing. This is seen from Fig. 2 where crystals A and C in the h phase continue to grow, whereas transformed crystals B and D stop

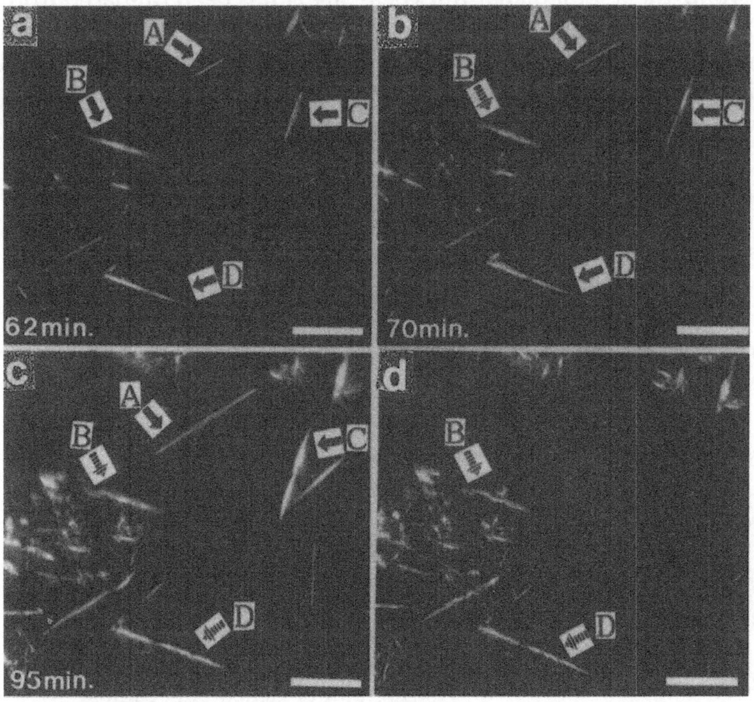

Fig. 2. In situ optical micrographs during isothermal and isobaric crystallization and melting (latter right). P = 2.82 kb, (ΔT) = 3°C. Scale bar = 50 μm

growing, at least on the time scale of the present observation.

The cross-sectional view of these crystals, as seen by transmission electron microscopy after etching and using two-stage replica technique, shows the tapered profile (Fig. 3). Since sharply fractionated PE (of chain length of about ~3000 Å) is used for experimentation, the tapered profile in Fig. 3 shows that, during the initial stages of crystallization, the chains are folded (~500 Å); they later thicken to full chain extension and even up to a few microns beyond. In the h phase thickening is feasible due to the high mobility of chains. Thus, in the h phase, simultaneous lateral and thickening growth of a crystal occurs (Fig. 3).

The mere fact that a new phase state first passes through a metastable state is, in itself, not quite unusual, and is expressed by the Ostwald Stage Rule dating back to 1898. With a reasonably appropriate choice of surface parameters [6] the situation can arise that the true thermodynamic stability conditions can invert with size. Specifically, for a PE single crystal which is sufficiently thin during the initial stages of crystallization, the h phase could be the thermodynamically stable phase. As the crystal thickens, above a critical value of lamellar thickness the initially stable h phase will no longer be thermodynamically stable. Thickening beyond the critical value of lamellar thickness will be due to kinetic reasons. Once a crystal transforms into the o phase, further thickening, along with lateral growth, becomes arrested or slows down drastically.

The fact that change in crystal size, in particular, thickening, can induce phase transformation has been successfully demonstrated on trans 1,4-polybutadiene (PTBD), Fig. 4 [8]. At ambient conditions, here, the crystals are in the monoclinic phase. With the increase of temperature, at atmospheric pressure, monoclinic to hexagonal transition occurs between 60°–85°C. This transition temperature is dependent on lamellar thickness. Chains in the h phase of PTBD are mobile, similar to that in the case of PE. The h phase here being accessible at atmospheric pressure makes PTBD an indeal case for this particular experimentation [8].

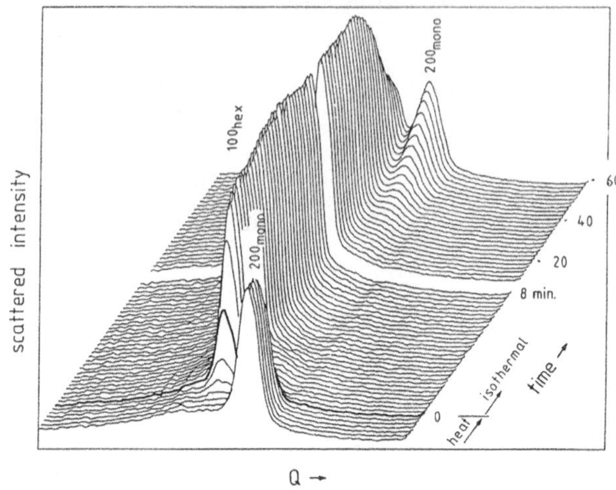

Fig. 4. Phase reversal of hexagonal to monoclinic as observed from the time-resolved WAXS

Figure 4 shows a time resolved WAXS pattern of PTBD. For the present purpose a single crystal mat of lamellar thickness 104 Å has been prepared from solution. For this thickness, monoclinic to hexagonal transition occurs at 68°C. The sample is left to anneal isothermally just above the monoclinic to hexagonal transition temperature. Once the crystals are in the h phase, they thicken. After certain isothermal annealing time, phase reversal from h to monoclinic sets in. This is indicative of the effect that, as the chains continuously extend, the initially thermodynamically stable h phase ceases to be the stable one and can undergo a hexagonal to monoclinic transition.

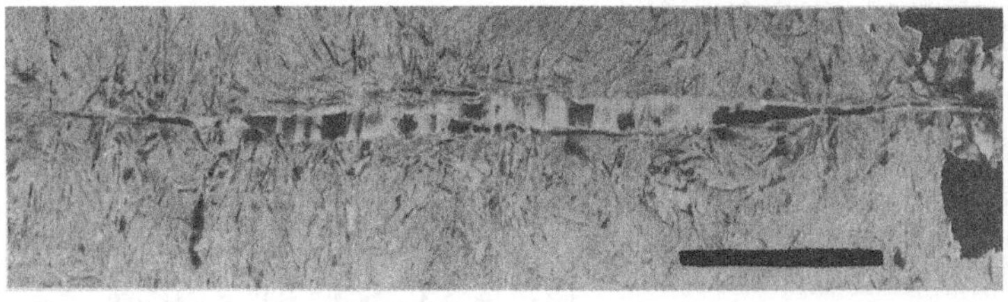

Fig. 3. Cross-sectional view of crystal, showing tapered edges. $P = 4.0$ kb, $(\Delta T) = 4°C$. Scale bar = 10 µm. An electron micrograph of etched bulk sample

References

1. Bassett DC, Khalifa BA, Turner B (1972) Nature 239:106; ibid 240:146
2. Bassett DC, Turner B (1974) Phil Mag 29:285; ibid 925
3. Wunderlich B, Melillo L (1968) Makromol Chem 118:250
4. Bassett DC (1977) High Temperatures-High Pressures 9:553
5. Hikosaka M, Rastogi S, Keller A, Kawabata H (1992) J Macromol Sci Phys 133:87
6. Rastogi S, Hikosaka M, Toda A, Keller A (in preparation)
7. Rastogi S, Hikosaka M, Kawabata H, Keller A (1991) Macromolecules 24:6384
8. Rastogi S, Ungar G (1992) Macromolecules (in press)

Authors' address:

Sanjay Rastogi
Physics Department
Lucknow University
Lucknow — 226007 (U.P.), India

Molecular dynamics in poly- and oligotetrafluoroethylenes

T. Albrecht[1]), H. Elben[1]), R. Jaeger[1]), M. Kimmig[1]), W. Petry[2]), C. Ritter[2]), H. Schwickert[3]), R. Steiner[4]), G. Strobl[1]), and B. Stühn[1])

[1]) Fakultät für Physik, Universität Freiburg, FRG
[2]) Institut Laue-Langevin, Grenoble, France
[3]) Hüls AG, Marl, FRG
[4]) Hatho GmbH, Freiburg, FRG

Abstract: Poly- and oligotetrafluoroethylenes exhibit at room temperature a high molecular mobility in the crystalline state. A detailed analysis of the molecular motion is possible by measuring the static and dynamic structure factor using x-ray and neutron scattering. The onset of diffusive molecular dynamics coincides in poly- and oligotetrafluoroethylenes with an order-disorder phase transition. — In PTFE the prevalent relaxational mode is a double helix reversal. Occurrence of helix reversals leads to a characteristic change in the static structure factor $S(q)$, indicating a loss of azimuthal correlation within the chain. A quasieelastic line in the dynamic structure factor $S(q, \omega)$ can be associated with motion of helix reversals. Six-fold symmetry (around the chain axis) of the intensity distribution of the diffusive scattering indicates an intermolecular correlation of rotational motion. — In $C_{20}F_{42}$ the phase transition at 200 K is mainly connected with the onset of rotational motion. Additionally, longitudinal diffuse motion sets in at the phase transition. This mode shows an increasingly collective character on approaching the phase transition.

Key words: Dynamics; PTFE; defects

Poly- and oligotetrafluoroethylenes exhibit, at room temperature, a high molecular mobility in the crystalline state. The state of motion is set up by confined longitudinal shifts, correlated rotations, and by mobile conformational defects ("helix reversals") (s. fig.). The onset of diffusive molecular dynamics coincides with an order-disorder phase transition. Details have been derived from q- and temperature dependent measurements of the static and dynamic structure factor, as obtained by x-ray- and neutron-scattering [1–3].

The occurrence of helix reversals becomes apparent by a change in the static structure factor; reversal defects lead to a loss of azimuthal correlation within the PTFE and $C_{20}F_{42}$ chain. Fitting the data to an appropriate theory yields an azimuthal correlation length in the range of 4–7 monomer units in the case of polytetrafluoroethylene (PTFE) and more than 9 monomer units for $C_{20}F_{42}$.

The prevalent relaxation mode in PTFE in an intermediate phase (292 K–303 K) is given by the motion of a double helix reversal (s. fig.). Since azimuthal long-range order is preserved, the corresponding Bragg reflections to not vanish. Diffusion of the reversal results only in local rotations of the chain.

The diffuse scattering in the disordered phase of $C_{20}F_{42}$ and PTFE exhibits a sixfold symmetry $S(\Phi)$ (Φ denotes the angle around the chain axis). This indicates an intermolecular correlation of the rotational dynamics.

The dynamic structure factor in the µeV range shows a quasieelastic line which is only present in the disordered phase. These dynamics can be interpreted assuming a reversal relaxation process and in the case of $C_{20}F_{42}$ also rotational diffusion. Data can be described by a stretched exponential for PTFE and by two lorentz lines for $C_{20}F_{42}$.

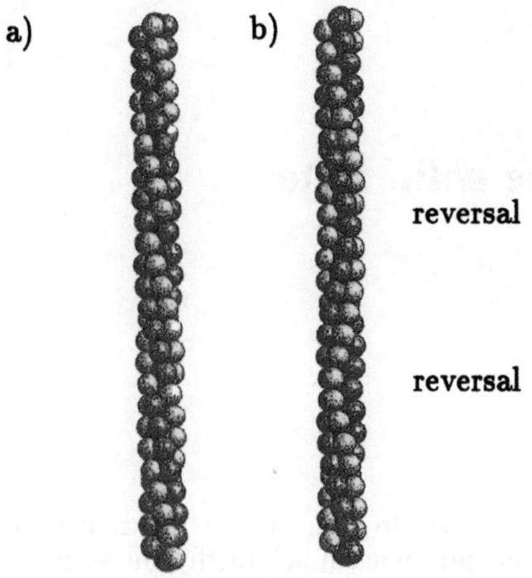

Fig. 1. a) PTFE-chain. b) A double helix reversal which changes the helicity of the PTFE-chain twice and, therefore, preserves the azimuthal long-range order

Additionally, the spectra exhibit broad lines in the meV-range which can be associated with librational and rotational vibrations. The broad-frequency distribution is caused by the angular disorder of the molecule chains and a damping due to interactions with other relaxation processes.

In $C_{20}F_{42}$ also a confined longitudinal diffusion is observed. Approaching the temperature of the order-disorder phase transition at $T = 200$ K this longitudinal diffusion gets an increasingly collective character. This is demonstrated by an increase of the intensity of the central line in the dynamic structure factor, and a simultaneous increase of the relaxation process time.

The dynamic structure factor also shows a longitudinal phonon. Dispersion relation and damping have been measured. Damping increased on approaching the Brillouin zone boundary. The phonons remain essentially unaffected by the order-disorder transition.

References

1. Schwickert H, Kimmig M, Strobl G (1991) J Chem Phys 95:2800—2806
2. Albrecht T, Elben H, Jaeger R, Kimmig M, Steiner R, Strobl G, Stühn B, Ritter C, Schwickert H (1991) J Chem Phys 95:2807—2816
3. Albrecht T, Jaeger R, Petry W, Steiner R, Strobl G, Stühn B (1991) J Chem Phys 95:2817—2822

Authors' address:

Prof. G. Strobl
Albert-Ludwigs-Universität
Fakultät für Physik
Hermann-Herder-Str. 3
D-7800 Freiburg, FRG

Overall mobility of polymethylene chains in the solid state: A spectroscopic study

M. Del Zoppo and G. Zerbi

Dipartimento di Chimica Industriale, Politecnico, Milano, Italy

Abstract: Vibrational spectroscopy is used to characterize the molecular phenomena involved in pretransitional processes in polymethylene systems. — It is shown that the "soliton-like" model proposed to justify the mechanism of molecular mobility is consistent with different experimental data from vibrational spectroscopy and other techniques, namely, diffusion, transport of matter, and surface disordering. — By use of spectroscopic "probes" such as Raman band width, frequency shifts, and intensity changes, it is predicted that, if solitons exist, we should observe in the IR/Raman spectra narrow band width and a possible upward shift of a few specific vibrational modes. Moreover, an intensity behavior different for distinct modes is predicted as a consequence of their different coupling with the torsional modes which, in turn are activated by the conformational soliton-like distortion of the chain. Vibrational spectra agree with what is predicted. — A test case (n-alkanes included in urea clathrate) is provided to support the predictions.

Key words: Mobility; polymethylene chains; spectroscopy

In dealing with polymethylene chains in the solid sate, one has to consider much experimental evidence which needs to be rationalized in a comprehensive and exhaustive framework. Here, we present the contributions which vibrational spectroscopy can provide in the elucidation of the molecular phenomena involved in the phase transitions considered.

We focus on the modifications which take place at the main solid-solid phase transition in n-alkanes. We refer to this phase as "α" phase. Its range of existence is chain-length- and pressure-dependent [1, 2].

It is well known that, upon heating, the crystalline lattice expands anisotropically, and the expansion coefficients have been experimentally determined [3, 4].

From IR and Raman spectroscopy we learn that at the crystalline low temperature (LT) to "α" transition the orthorhombic lattice disappears while the chains keep their substantially transplanar conformation [5]. Moreover, the conformational disorder which sets in at the transition is located mainly at chain ends [1, 5, 6].

A detailed analysis of C_{19} and careful measurements of IR intensities have shown that in the "α" phase a sizeable fraction ($\approx$25% of the chains) slides out of the lamellae and exhibits conformational disorder in positions 2—3 and 3—4 [5]. This suggests that, not only surface disordering takes place, but, also, longitudinal diffusion is likely to occur.

Diffusion either longitudinal or transversal is further proven by IR and DSC experiments in which, upon annealing, a mechanical mixture of n-alkanes of slightly different chain length turns into a cocrystal [7, 8].

Evidence that such a diffusion is longitudinal comes from neutron-scattering experiments [9, 10].

We look for a model which can take into account all these phenomena.

Vibrational spectroscopy provides a body of information which can be exploited to study the molecular nature of these phenomena. From the analysis of a spectrum, one obtains three kinds of information, namely: peak frequency, bandwidth, and intensity. It is known that the Raman bandwidth is a sensitive probe of the libro-torsional mobility of polymethylene systems [11—13]. Moreover, the combined analysis of band shape and peak frequency allows to distinguish between librational, rigid, rod-like, and purely torsional motions.

Indeed, the analysis of the Raman spectra in the C—H stretching region reveals that [13]:

i) if chains perform large amplitude overall rotations their motion should be detected from bandwidth of d^-;
ii) torsional motions can also contribute to the broadening of d^-;
iii) if the chains become torsionally distorted the frequency of d^- shifts upward.

The coupling of d^- with torsional modes, which is the origin of the upward frequency shift, is proven both by normal coordinate [13] and quantum mechanical, "ab initio" calculations [14].

These findings are nicely verified in the spectra of n-alkanes included in urea channels [13]. In this case, we know that chains are forced to keep their transplanar conformation (there is not enough room for the formation of G defects), but are free to perform large amplitude libro-torsions because of the weak intermolecular interactions.

Among the different models which have been proposed, the one we feel is in better agreement with the experimental evidence is the soliton-like model first proposed by Mansfield [15]. According to this model, a slight collective distorsion of the C—C skeleton can propagate through the lattice and in so doing causes transport of matter and, hence, diffusion.

From what was discussed above, if such twistons do exist and are at the origin of the molecular phenomena involved, we would predict in the IR/Raman spectra narrow bandwidth and a possible upward shift of d^- if an appreciable torsional coupling is allowed.

In addition, if at the LT → "α" transition the system evolves as a whole, we should be able to detect its evolution from some property which reflects the global behavior of the system. An appealing quantity is IR intensity.

If a harmonic and/or anharmonic coupling of the high-frequency modes with the low-frequency torsional modes becomes effective upon conformational distortion of the chain, we would predict transfer of intensity from the high-frequency modes to the torsional ones. The coupling is stronger the lower the frequencies of the internal modes, thus suggesting different intensity losses in the distinct regions of the spectrum. Indeed, the IR spectra, starting already from very low temperatures, show a remarkable decrease in intensity, as reported by Snyder [16]. These intensity changes can hardly be explained if the concept of twistons is not introduced [17].

Thus, the twiston model seems to be consistent with the large collection of observed spectroscopic data in the "α" phase and can explain transport of matter, diffusion, and surface disordering. In summary, we can say that the spectroscopic experimental data can be explained according to the following observations:

i) Disappearance of orthorhombic lattice: chains are out of register because of twiston diffusion.
ii) Uneven distribution of conformational disorder inside the chain: gauche defects are located mainly at chain ends as a consequence of diffusion.
iii) Narrow bandwidth: no large amplitude librational motions are allowed.
iv) Possible upward shift: the amount of frequency shift depends on the extent of torsional distorsion, being sizeable for systems in a "softer" surrounding (e.g., inclusion compounds [13], solid aqueous solution of decyl-ammonium chloride [14]).
v) Intensity changes: consequence of the coupling of high-frequency modes with torsional modes.

Conclusions

We propose a mechanism of molecular mobility for polymethylene chains in the solid state which justifies, in a comprehensive way, much independent experimental data from vibrational spectroscopy or from other experiments.

The model requires further critical verification with other experimental or theoretical techniques. We wish to point out that if NMR experiments will be used it should be kept in mind that the frequencies of the torsional vibrations eventually implied are of the order of 10^{12}—10^{14} Hz.

At present, numerical simulations based on molecular dynamics calculations seem to be one way to delve into molecular motions. The situation seems to us to be very similar to that of many years ago when computers allowed to carry out normal-mode calculations on extremely large systems. The main problem in both cases is the reliability of the inter and intramolecular potentials used. In our experience, the results are always strongly force-field-dependent.

The challenge presented to molecular simulations by vibrational spectroscopy in the search for twistons is the fact that such numerical simulations must account, not only for frequencies, but also for band shapes and band intensities. In other words, intramolecular charge flows, anharmonic coupling and other quantum effects should be suitably accounted for.

References

1. Maroncelli M, Strauss HL, Snyder RG (1985) J Chem Phys 82:2811–2824
2. Ungar G (1986) Macromolecules 19:1317–1324
3. Strobl GR (1977) J Pol Sci: Pol Symp 59:121–142
4. Ungar G (1983) J Phys Chem 87:689–695
5. Zerbi G, Magni R, Gussoni M, Moritz KH, Bigotto A, Dirlikov S (1981) J Chem Phys 75:3175–3194
6. Del Zoppo M, Zerbi G (1990) Polymer 31:658–662
7. Zerbi G, Piazza R, Moritz KH (1982) Polymer 23:1921–1928
8. Ungar G, Keller A (1979) Colloid Polym Sci 257:90–94
9. Ewen B, Richter D (1978) J Chem Phys 69:2954–2963
10. Guillaume F, Doucet J, Sourrisseau C, Dianoux AJ (1989) J Chem Phys 91:2555–2567
11. Rothschild WG (1984) "Dynamics of Molecular Liquids" Wiley, New York
12. Wunder SL, Bell MI, Zerbi G (1986) J Chem Phys 85:3827–3839
13. Zerbi G, Roncone P, Longhi G, Wunder SL (1988) J Chem Phys 89:166–173
14. Del Zoppo M, Zerbi G, Polymer (in press)
15. Mansfield M, Boyd RH (1978) J Pol Sci: Pol Phys Ed 16:1227–1252
16. Snyder RG, Maroncelli M, Strauss HL, Hallmark VM (1986) J Phys Chem 90:5623–5630
17. Zerbi G, Del Zoppo M to appear in Faraday Symposium on "The Conformation of Flexible Molecules in Fluid Phases" (1992) J Chem Soc Far Trans 88

Authors' address:

Prof. G. Zerbi
Dip. di Chimica Industriale
Politecnico di Milano
Piazza Leonardo da Vinci, 32
20133 Milano, Italy

Liquid crystalline polymers

Hexagonal columnar phases in polymers

G. Ungar

School of Materials, University of Sheffield, UK

Abstract: Polymeric mesophases with positional order in two dimensions are considered in relation to the established liquid crystal phase classification. It is argued that the mesophase with hexagonal symmetry, observed in polymers as flexible as polyethylene and poly(diethylsiloxane), is a columnar liquid crystal. This phase, well known from discotic low molecular compounds, is shown to be present in a wide variety of non-discotic polymers. Some recent developements involving aperiodic main-chain and hemiphasmid side-chain polymers are summarized.

Key words: Columnar liquid crystals; hexagonal phase; liquid crystal polymers; two-dimensional order

Introduction

Practically the entire past and present research on liquid crystal polymers LCPs involved either nematic or smectic phases. As is well known, nematic liquid crystals possess only orientational long-range order, whilst smectics have positional LRO (or, strictly, quasi-long-range order, QLRO [1]) in only one dimension. It has remained largely unrecognised that there is a considerable number of polymers exhibiting the third liquid crystal type, the columnar phases [2, 3, 4], otherwise well known to form by low molar mass disc-like molecules [5]. The columnar phase has true LRO in two dimensions, but not in the third. Table 1 summarises the updated, currently accepted, symmetry-based mesophase classification, including the limiting cases of crystal and isotropic liquid (cf. also [6, 7]).

Unlike the discotic phases, in the prevailing non-discotic polymeric columnar phase molecules extend along the columns. X-ray fibre patterns show sharp equatorial reflections indexable on a hexagonal lattice, while only diffuse off-equatorial scattering is observed. Polymeric columnar phase is unique in the whole field of liquid crystals in that it is formed by molecules which often possess neither rigid units nor amphiphilic character.

In terms of their constitution, many column-forming polymers do not resemble traditional LCPs. Different types of macromolecules exhibiting the non-discotic columnar phase are compiled in Table 2. These can be grouped into A) Linear flexible or semiflexible, B) Branched with flexible backbone and C) Branched with rigid backbone.

Linear flexible or semiflexible polymers

The representatives of this first group include such unconventional LCPs as 1) polyethylene — the "hexagonal phase" induced by high pressure [8, 9], irradiation [10] or ultradrawing [11]; 2) poly-1,4-t-butadiene [12]; or 3) a number of halogenated ethylene polymers.

However, the columnar phase has recently been reported in main chain copolyethers that resemble more closely the conventional LCP's [13]: they comprise "mesogenic" and "spacer" moieties and also exhibit smetic phases. The "mesogen" is either biphenylethane (BPE) [13] or fluorobiphenylethane (FBPE) [14], while flexible aliphatic spacers of two different lengths are distributed randomly along the chain. The resulting chemical aperiodicity appears to be responsible for the absence of positional LRO along the chain axis, thus favouring 2-D (columnar) over 3-D (crystalline) ordering.

Table 1. Phase classification in terms of translational symmetry

	Dimensionality of long-range order			Phase
	Mol. Orient. Order*)	Bond Orient. Order*)	Positional Order	
Isotropic liquid	0	0	0	Iso.
		Liquid crystal		
Nematic	1	0	0	N_u (uniaxial)
	3	0	0	N_b (biaxial)
Smectic	1	0	1**)	Sm-A
	3	2	1**)	Sm-C
	3	2	1**)	Hexatic phases (Hexatic-B, Sm-F, Sm-I)
Columnar	3	2	2	Columnar
		Crystal		
Disordered crystal	3	3	3	"Ordered Smectics"-crystal B, G, J, E, H, K "Smectic-D" (cubic) Plastic crystal Other orientationally and conformationally disordered crystals
Ordered crystal	3	3	3	Ordered crystal

*) Note the distinction between molecular orientational and bond orientational long-range order (LRO). The former refers to the orientation of a tensorial quantity associated with the molecule itself, like moment of inertia or polarizability. The latter refers to the orientation of the vector joining two neighbouring molecules.

**) Quasi-long-range order (QLRO) (In QLRO correlation decays algebraically, and asymptotically approaches zero with increasing distance; in LRO it descends to a finite value.)

In the case of the columnar phase of chemically periodic flexible polymers of group A (polyethylene etc.), the necessary aperiodicity is believed to be provided by the thermally activated conformational disorder (see e.g. the paper by Wunderlich and coworkers, this volume [15]).

Branched polymers with flexible backbone

Branched polymers with flexible backbones (B) include n-alkyl substituted polysiloxanes and polysilanes, as well as substituted polyphosphazenes (Table 2). The branches can be as short as C_2H_5 in poly(diethylsiloxane), making the distinction between polymer classes A and B (Table 2) somewhat arbitrary.

Recently, polymers with forked [16] and hemi-phasmid [17, 18] side groups were also shown to form the hexagonal columnar phase. A typical representative of either polymer type is shown below.

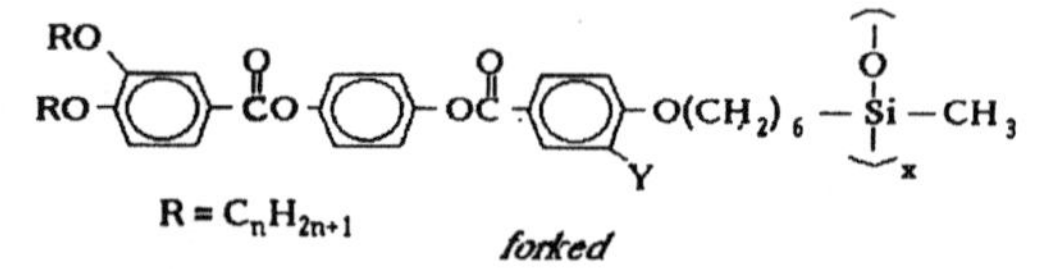

forked

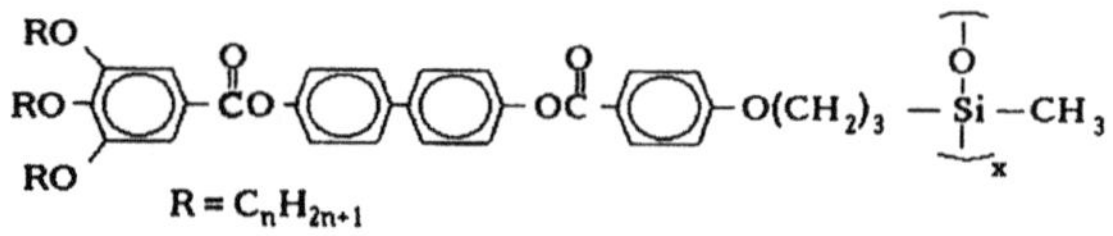

hemi-phasmid

It was found [17] that longer aliphatic tails favour the columnar phase, while shorter tails favour smectic ordering. The tendency toward column formation may be attributed to steric overcrowding of the molten tails and the generally tapered shape of the mesogens. Diffraction data are consistent with the model in which side groups radiate out from the polymer backbone located in the centre of the column (see Fig. 1).

Interesting re-entrant effects are observed in some hemi-phasmid columnar polymers, with the first case of re-entrant isotropic phase in a LCP being discovered recently [18].

Branched polymers with rigid backbone

The only clear, reported representatives of this type showing hexagonal columnar phase are a series of synthetic polypeptides, poly(n-alkylglutamates) [19, 20], where the periodic potential of the rigid helical backbone is apparently screened off at sufficiently high temperatures by the molten aliphatic sheath; hence the chains effectively act as smooth cylinders and pack on a hexagonal lattice. Similarly, as with hemi-phasmid polymers, re-entrant behaviour is sometimes observed; in this case, cholesteric phase occurring below the columnar phase [20].

Table 2. Non-discotic polymers exhibiting a thermotropic 2-D hexagonal phase

Polymer	Formula	Common name	Temperature interval (°C)	Ref.
A)				
— polyethylene				
— at high pressure (triple point 3.5 kbar, 215 °C)		hexagonal, anabaric	280—307 (@ 8 kbar)	[8, 9]
— constrained ultra-oriented fibres		hexagonal	155—180	[11]
— irradiated		hexagonal	60:130—100:135*)	[10]
— PE with short branches:				
ethylene-propylene copolymer		hexagonal	<60	[10]
ethylene-vinylchloride copolymer		hexagonal	<20:54	[21]
— 1,4-t-polybutadiene	$(—CH_2—CH{=}CH—CH_2—)_x$	Form II	83—164	[12]
— polytetrafluoroethylene	$(—DF_2—CF_2—)_x$	Form I	30—330	[22, 23]
— tetrafluoroethylene-hexafluoropropylene copolymer			>—40	[24]
— polytrifluoroethylene	$(—CF_2—CHF—)_x$		<r.t. 4—193:209	[25, 26]
— vinylydenefluoride/trifluoroethylene copolymer (50—80% VDF)		paralelectric phase	70:150—150:160	[27]
— poly-p-xylylene	$(—Ph—CH_2CH_2—)_x$	β_2	280—420	[28]
— polyethers of biphenylethane (BPE-m,n):	$(—Ph—CH_2CH_2—Ph—O—(CH_2)_{m,n}—O—)_x$,	columnar	104—181	[13]
fluorobiphenylethene (FBPE-m,n):	$(—Ph—CH_2CH_2—Ph(F)—O—(CH_2)_{m,n}—O—)_x$,		110—140	[14]
and BPE-FBPE copolymers			100—180	[14]
B)				
— polydialkylsiloxanes, R = C_2H_5, C_3H_7	$(—SiR_2—O—)_x$	α_m, μ	—50: +40—50:200	[29]
— polyphosphazanes, R = $F_3C—CH_2$ or aromatic	$(—P(OR)_2{=}N—)_x$	δ	150:180—360:400	[30, 31, 32]
— polydialkylsilanes, R = n-C_4H_9 through n-$C_{14}H_{29}$	$(—SiR_2—SiR_2—)_x$	Phase II, columnar	10:80—>450	[33, 34]
— polydialkylgermanes, R = n-C_6H_{13}	$(—GeR_2—GeR_2—)_x$		>15	[35]

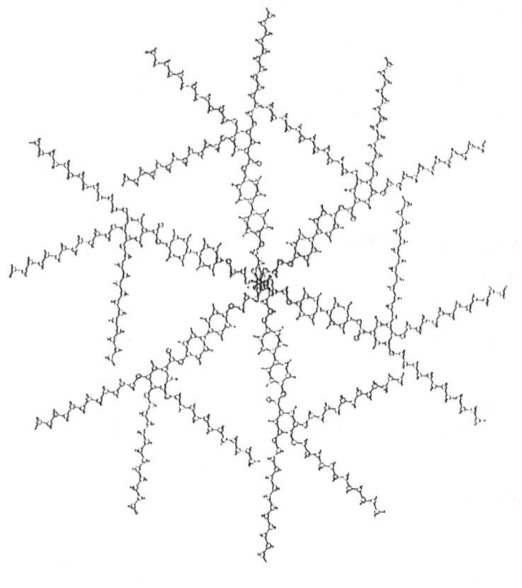

Fig. 1. Schematic view along the column axis of a hemiphasmid side-chain polymer. The real structure is more disordered, with n-alkyl tails having non-extended conformations

Conclusion

A sizeable proportion of polymers displays a partially ordered phase with two-dimensional positional LRO and hexagonal symmetry. Notwithstanding the fact that its physical properties, such as heat of isotropisation or viscosity, can vary considerably from one polymer to another, this phase is properly described as hexagonal columnar liquid crystal. Recent work has provided a clearer understanding of design principles involved in tailoring new column-forming polymers, including those with anticipated application potential.

References

1. Caille A (1972) Compt Rend Acad Sci Paris 274B:891
2. Weber P, Guillon D, Skoulios A, Miller RD (1989) J Phys 50:793
3. Ungar G (1989) Proc Biennial Conf Polym Phys Group, Inst Phys, Reading (submitted to Polymer)
4. Ungar G, Feijoo JL (1990) Mol Cryst Liq Cryst 180B:281
5. Destrade C, Foucher P, Gasparoux H, Tinh NH and Levelut AM (1984) Mol Cryst Liq Cryst 106:121
6. Pershan PS (1988) Structure of Liquid Crystal Phases. World Scientific, Singapore
7. Leadbetter AJ (1987) in: Gray GW (ed) Thermotropic Liquid Crystals, Wiley, New York
8. Bassett DC, Turner B (1972) Nature, Phys Sci 240:146; Bassett DC, Block S, Piermarini G (1974) J Appl Phys 45:4146
9. Yamamoto T (1979) J Macromol Sci — Phys B16:487
10. Ungar G, Keller A (1980) Polymer 21:1273
11. Pennings AJ, Zwijnenburg A (1979) J Polym Sci, Polym Phys Ed 17:1011
12. Suehiro K, Takayanagi M (1970) J Macromol Sci — Phys B4:39
13. Ungar G, Feijoo JL, Percec V, Yourd R (1991) Macromolecules 24:953
14. Percec V, Zuber M, Ungar G, Alvarez-Castillo A (1992) Macromolecules 25:1193
15. Wunderlich B, this volume
16. Achard MF, Nguyen Huu Tinh, Richard H, Mauzac M, Hardouin F (1990) Liq Cryst 8:533
17. Percec V, Heck J, Ungar G (1991) Macromolecules 24:4957
18. Percec V, Heck J, Blackwell H, Ungar G, Alvarez-Castillo A J Mater Chem, in press
19. Watanabe J, Ono H, Uematsu I, Abe A (1985) Macromolecules 18:2141
20. Watanabe J, Takashina Y (1991) Macromolecules 24:3423
21. Gomez MA, Tonelli AE, Lovinger AJ, Schilling FC, Cozine MH, Davis DD (1989) Polymer Prepts Am Chem Soc 30:317
22. Clark ES, Muus LT (1962) Z Krist 117:119; Clark ES (1967) J Macromol Sci — Phys B1:795
23. Corradini P, Guerra G (1977) Macromolecules 10:1410
24. Weeks JJ, Eby RK, Clark ES (1981) Polymer 22:1496
25. Galperin YL, Strogalin YuV (1965) Polymer Sci USSR 7:15
26. Lovinger AJ, Cais RE (1984) Macromolecules 17:1939
27. Lovinger AJ, Furukawa T, Davis GT, Broadhurst MG (1983) Polymer 24:1225, 1233
28. Isoda S, Kawaguchi A, Katayama K-I (1984) J Polym Sci, Polym Phys Ed 22:669
29. Godovsky YuK, Papkov VS (1989) Adv Polym Sci 88:129
30. Allen G, Lewis CJ, Todd SM (1970) Polymer 11:44
31. Schneider NS, Desper CR, Beres JJ (1978) in: Blumstein A (ed) Liquid Crystalline Order in Polymers, Academic Press, New York
32. Kojima M, Magill JH (1989) Polymer 30:579
33. Lovinger AJ, Schilling FC, Bovey FA, Zeigler JM (1986) Macromolecules 19:2660
34. Weber P, Guillon D, Skoulios A, Miller RD (1990) Liq Cryst 8:825
35. Patnaik SS, Farmer BL (1990) Polym Prep, Amer Chem Soc 31(2):292

Author's address:

Dr. Goran Ungar
School of Materials
University of Sheffield
Sir Robert Hadfield Building
Mappin Street
Sheffield S1 3JD, England

Healing of knit lines in injection-molded liquid crystalline polymers

K. Engberg and U. W. Gedde

Department of Polymer Technology, Royal Institute of Technology, Stockholm, Sweden

Abstract: Randomization and healing of oriented layers and knit line regions in injection-molded samples of thermotropic Vectra A950 [poly(4-hydroxy benzoic acid-co-2-hydroxy-6-naphthoic acid)] have been studied by thermal optical analysis (TOA), thermal mechanical analysis (TMA), differential scanning calorimetry (DSC), wide-angle x-ray scattering (WAXS), and tensile testing. Annealing below the crystal melting temperature range resulted in no healing effects, whereas the knit line strength increased progressively with increasing annealing temperature within the melting range 275–293 °C. TOA, TMA, WAXS, and DSC on oriented samples taken from the surface layer showed that oriented films undergo a randomization process when annealed in the melting temperature range. The macroscopic chain orientation decreased suddenly on heating in this temperature range and appeared also to be constant under isothermal conditions. It is proposed that the melting of small crystallites is a prerequisite for this randomization. Isothermal annealing of the fully nematic melt caused a very slow randomization of the oriented structure.

Key words: Liquid crystalline polymers; orientation; knit lines; healing; crystal melting

Introduction

The mechanical strength of knit lines formed in thermotropic liquid crystalline (LC) polymers is low [1]. Welds formed by a co-linear impingement of two melt fronts exhibited a fracture stress of 10.8 ± 1.6 MPa which corresponded to 15–20% of the full strength of the material [1]. It was shown that a 200-µm-thick region with moderate transverse orientation was formed on both sides of the welds [1]. Some annealing experiments on Vectra samples with knit lines were also reported and a limited degree of healing was observed in samples annealed at temperatures higher than 280 °C [1]. The present paper presents more extensive data on annealing of samples with cold knit lines. The knit line healing process can be thought of as consisting of the establishment of physical contact between the two surfces, the diffusion of molecular segments across the phase boundary and, finally, the formation of new entanglements in the interface region [2]. The knit line healing in LC polymers was believed to be associated with randomization of the oriented interface followed by chain segment diffusion and the formation of chain entanglements. This paper critically compares data relating to knit line strength of annealed samples with data from thermal analysis and x-ray diffraction recording the randomization of oriented LC polymer samples.

Experimental

A liquid crystal copolyester, Vectra A950 (Hoechst) prepared from 4-hydroxy benzoic acid and 2-hydroxy-6-naphthoic acid, was used in this study. Samples were prepared by injection-molding using a Netsal Neomat 70/75 equipped with two molds. The pellets were dried at 160 °C for 4 h prior to the processing. The first mold generated ruler-shaped samples with dimensions 100 × 20 × 2 mm containing a knit line formed by a colinear impingement of the two melt fronts. The melt flow directions were parallel to the long side of the rulers. The second mold generated dumbell-shaped samples according to DIN 53 455. The following processing conditions were used: melt temperature: 320°C, mold temperature: 100°C, injection time: 2.0 s (0.2 s for the DIN samples), and holding pressure: 3.5 MPa.

The samples with knit lines were annealed in a specially designed mold for 5 min under isothermal conditions at temperatures between 160° and 300°C. All samples were dried at 160°C for 4 h prior to the annealing. A small compression load was applied over the welding area during the heat treatment.

The fracture stress of the ruler-shaped samples was obtained in an Instron Universal Testing Instrument Model 1122 at 25°C and at an elongation rate of 1 mm min^{-1}. At least five samples were tested for each material. The stress was applied parallel to the melt flow direction. The microstructure of the knit line region was studied by scanning electron microscopy (SEM) on samples fractured at liquid nitrogen temperature. After being coated with Pt-Pd, the samples were examined in an ISI Super Mini SEM.

A number of reference oriented samples were prepared by machining the 100-μm top layer from the DIN samples. X-ray diffraction showed that these samples were oriented with the following values of Hermans orientation function: $f_{c,c} = 0.79$ (crystalline component) and $f_{c,a} = 0.69$ (amorphous component). Thermal optical analysis (TOA), using a Leitz Ortolux POL BK II equipped with crossed polarizers and a temperature calibrated Leitz hot-stage, was performed on 2-μm samples sectioned from the 100-μm-thick top layer. The samples were adjusted in the hot stage so that the optical director was at an angle of 45° to the crossed polarizers of the microscope. The intensity of the light passing through the column was measured during the annealing in the temperature range 200°–320°C. The oriented 100-μm samples were first annealed in a temperature-calibrated Perkin-Elmer DSC-7 under isothermal conditions in the temperature range 270°–310°C, and then studied by wide-angle x-ray scattering (WAXS) using a STADI/P instrument equipped with a curved position-sensitive detector with a radius of 220 mm. Strictly monochromatized CuKα_1 radiation with $\lambda = 0.1540560$ nm was used. Diffractograms were obtained from the samples at different angular positions with respect to a fixed director in order to obtain the azimuthal-angle-dependence of the scattering in order to determine the chain-axis orientation.

The chain-axis orientation was calculated from the x-ray diffraction pattern according to Eqs. (1–3). The director was set parallel to the knit line and coincided with the chain axis director. The x-ray diffraction pattern was composed of two major reflections corresponding to d-spacing values 0.21 nm (meridional) and 0.45 nm (equatorial) [3]. The equatorial scattering consisted of a Bragg reflection at $2\theta \approx 19.5°$ (d = 0.45 nm) and a broad amorphous halo peaking at $2\theta \approx 19.5°$ and 27° (Fig. 1). The 0.45 nm Bragg reflection and the "equatorial" amorphous halo were used to obtain the chain-axis orientation of the crystalline and amorphous components (Eqs. (1–3)):

$$\langle \cos^2 \phi_{\mathrm{inter},i} \rangle = \frac{\sum_{\phi=0°}^{180°} I_{\phi\mathrm{inter},i} \sin\phi \cos^2\phi}{\sum_{\phi=0°}^{180°} I_{\phi\mathrm{inter}} \sin\phi} \tag{1}$$

$$f_{\mathrm{inter},i} = \frac{3\langle \cos^2 \phi_{\mathrm{inter},i} \rangle - 1}{2} \tag{2}$$

$$f_{c,i} = -2\, f_{\mathrm{inter},i}\ , \tag{3}$$

where ϕ is the azimuthal angle, $I_{\phi\mathrm{inter},i}$ is the intensity of scattering associated either with the Bragg reflection ($i = c$) or with the amorphous halo ($i = a$), $f_{\mathrm{inter},i}$ is the Hermans orientation [4, 5] function of the interchain spacing (crystalline (c) or amorphous (a)), and $f_{c,i}$ is the chain axis orientation associated with the crystalline or amorphous component. Equation (3) was obtained under the assumption that orientation is uniaxial. This assumption was supported by the absence of Bragg reflections at $\theta = 0°$. The standard deviation of a single data point was determined based on four repeated and independent x-ray measurements to 0.012 (in $f_{c,c}$; average value is equal to 0.804) and 0.010 (in $f_{c,a}$; average value is equal to 0.691).

Crystal melting was recorded in the standard oriented sample and in the standard isotropic sample by DSC (Perkin-Elmer DSC-7) and TMA (Perkin-Elmer TMS-2), in both cases at a heating rate of 20°C min^{-1}. The standard isotropic sample was prepared from a near-isotropic sample cut from the core of the DIN sample by further annealing at 340°C for 0.5 min. TMA allowed the measurements of thermal expansion at zero load in a direction perpendicular to the direction of orientation and the penetration of a loaded (50 mN) probe.

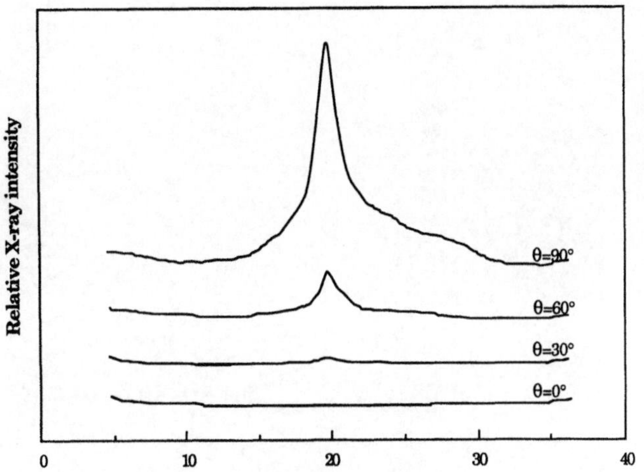

Fig. 1. X-ray diffraction patterns for an oriented reference sample obtained at different azimuthal angles as indicated in the figure. The equatorial scattering (ϕ = 90°) is used to illustrate the sharp crystalline reflection and the amorphous halo

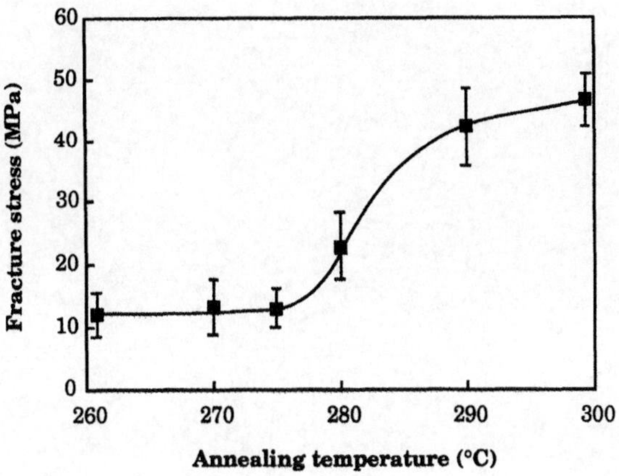

Fig. 2. Fracture stress of annealed samples as a function of annealing temperature. Bars indicate the standard deviation of a single data point

Results and discussion

Figure 2 shows the strength of cold knit lines in annealed samples as a function of annealing temperature. All fractures were confined to the knit lines and fracture occurred without yielding. Samples annealed at temperatures lower than 275°C exhibited strength values similar to those previously reported for unannealed samples, i.e., 10.8 ± 1.6 MPa [1]. A strong increase in strength was observed with increasing annealing temperature in the temperature range 275°—290°C. Samples annealed at even higher temperatures exhibited only a small increase in knit line strength with increasing annealing temperature. The maximum strength value was obtained for samples annealed at 300°C and amounted to 46.8 ± 4.2 MPa. This is a fourfold increase in strength compared to that of the unannealed samples. Samples annealed in the melt temperature range for a longer period of time, i.e., 10 min, exhibited approximately the same knit line strength as those annealed for a shorter period of time, i.e., 5 min.

The scanning electron micrographs presented in Fig. 3 further substantiate the effects noted in Fig. 2, namely, that annealing at low temperatures caused no change in orientation of the near-knit line material, whereas annealing at higher temperatures led to a signficant reduction of chain orientation. The original fibrous character of the near-knit line material was not present in the sample which was annealed at 290°C (Fig. 3c).

Figure 4 presents DSC and TMA thermograms of the reference oriented and the reference isotropic samples, and it may be concluded that the solid-nematic transition (melting) involved very small energies: 0.4—0.9 kJ kg^{-1} (oriented) and 1.6—2.4 kJ kg^{-1} (isotropic). The melting range of the oriented sample was 278—300°C, with two peaks appearing at 282° and 294°C. The isotropic sample exhibited a unimodal melting ranging from 260° to 290°C. The penetration data shown in the same figure, indicating softening at 270°C, is in fair agreement with the data obtained by DSC. It may be concluded that melting is affected by the state of orientation of the sample. The material adjacent to the cold welds exhibited a degree of orientation of $f_{c,c}$ = 0.213 and $f_{c,a}$ = 0.206, which is less than that of the reference oriented sample. It can be expected that the solid-nematic transition should occur at lower temperatures in the near-knit line material than in the reference oriented sample. The range of crystal melting in the near-knit line material is estimated to be 270 ± 10°C to 295 ± 5°C.

The randomization of the reference oriented polymer was recorded by TOA, TMA, and WAXS (Figs. 5 and 6). The transmitted light intensity (curve a in Fig. 5) was constant up to 278°C, indicating that no change in chain orientation occur-

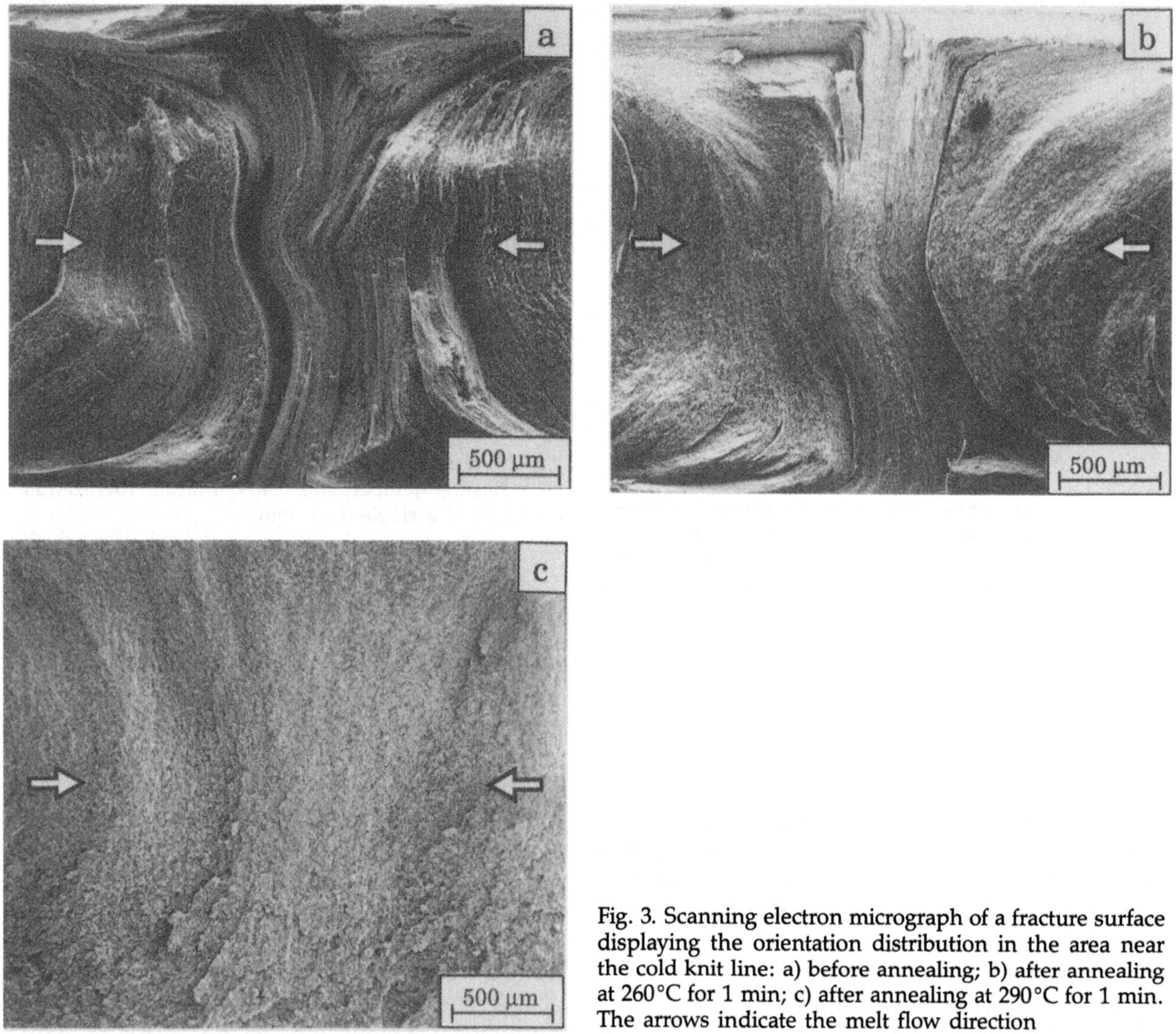

Fig. 3. Scanning electron micrograph of a fracture surface displaying the orientation distribution in the area near the cold knit line: a) before annealing; b) after annealing at 260 °C for 1 min; c) after annealing at 290 °C for 1 min. The arrows indicate the melt flow direction

red in the sample at these temperatures. A strong decrease in transmitted light intensity with increasing temperature was observed in the temperature range 278°–293 °C, whereas at even higher temperatures no appreciable change in light intensity was observed. The results showed no dependence on heating rate at rates between 1° and 20 °C min^{-1}. The data presented in curves b–d of Fig. 5 confirm the aforementioned findings that no reorientation occurred at low temperatures and that rapid reorientation (randomization) occurred at temperatures between 278° and 293 °C. The crystalline orientation is always greater than the amorphous orientation, cf. curves b and c in Fig. 5. However, as shown in the next figure, most of the randomization occurred instantaneously. The diffusive process which rendered the randomization occurred faster than the minimum time for measurement.

Figure 6 presents the effect of interrupting the constant-rate heating, holding the temperature constant for a period of time, and then, finally, again heating the sample at constant rate. TOA and TMA did not allow the samples to establish thermal equilibrium in the initial part of the isothermal phase. The change in both linear expansion and light intensity recorded in the initial part of the isothermal phase is most probably due to this

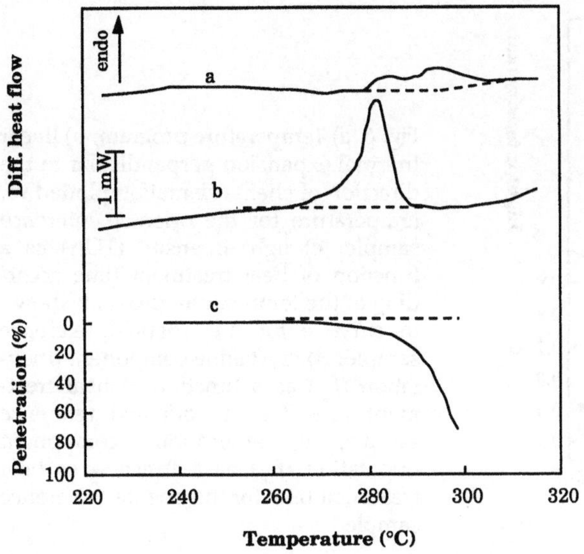

Fig. 4. a) Melting thermogram of oriented reference sample of Vectra ($f_{c,c}$ = 0.79 and $f_{c,a}$ = 0.69); b) melting thermogram of isotropic sample; c) TMA data showing penetration as a function of temperature of a sample consisting of an oriented surface layer ($f_{c,c}$ = 0.79 and $f_{c,a}$ = 0.69) and an isotropic core. The load applied to the sample was 50 mN

thermal non-equilibrium. The temperature control for the samples used in the x-ray diffraction experiments was much superior to that used in TOA and TMA and it is shown by WAXS that chain orientation remains constant during the isothermal phase. This was further checked by annealing a sample for 60 min at 285 °C and, indeed, no change in chain orientation was observed. The orientation in this ample ($f_{c,c}$ = 0.68 and $f_{c,a}$ = 0.55) was essentially the same as that in the samples annealed for shorter periods of time, i.e., for 0.1 and 5 min, exhibiting the orientation values: $f_{c,c}$ = 0.66; $f_{c,a}$ = 0.57, and $f_{c,c}$ = 0.65; $f_{c,a}$ = 0.50). Remaining chain orientation was always observed in samples at temperatures greater than 295 °C. A very slow randomization process also occurred at the highest temperatures: annealing at 310 °C for 30 min caused a decrease in chain orientation from an f-value of 0.50 (0.42) to a value of 0.24 (0.16). The first values are associated with crystalline component orientation and the values within brackets are values for the amorphous phase orientation.

The chain orientation of the crystalline phase was always greater than for the amorphous phase, which is consonant with the condition that chain orientation of an arbitrary small structural region (with reference to the director of the region) is always greater within crystals than within amorphous regions of corresponding size.

The near-knit-line material exhibited a smaller chain orientation than the reference oriented material. However, the observations made for the reference oriented sample are applicable on the near-knit-line material. Annealing at 260 °C for 1 min of the near-knit-line sample caused no change in orientation; $f_{c,c}$ and $f_{c,a}$ remained constant at 0.215 and 0.200. Annealing at 290 °C for 1 min

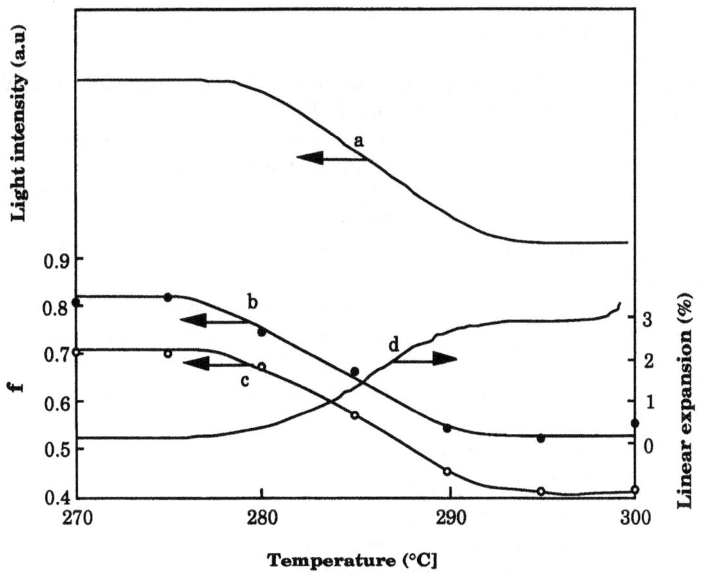

Fig. 5. a) Transmitted light intensity (TOA) as a function of temperature for the oriented reference sample. Temperature scanning rate: 5 °C/min; b) crystalline component orientation ($f_{c,c}$) as a function of annealing temperature for oriented reference sample annealed at each temperature for 0.1 min; c) amorphous component orientation ($f_{c,a}$) as a function of annealing temperature for oriented reference sample annealed at each temperature for 0.1 min; d) linear expansion of the oriented reference sample as a function of temperature

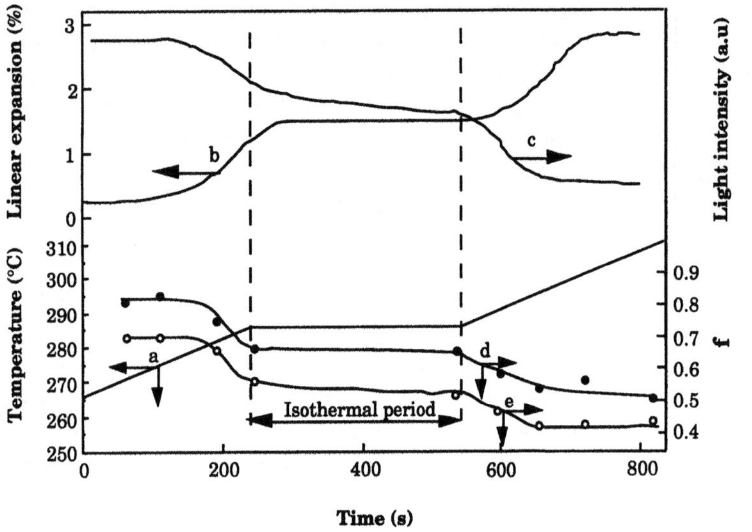

Fig. 6. a) Temperature program; b) linear thermal expansion perpendicular to the direction of chain orientation plotted vs. temperature for the oriented reference sample; c) light intensity (TOA) as a function of heat treatment time according to the temperature program shown in curve a for the oriented reference sample; d) crystalline component orientation ($f_{c,c}$) as a function of heat treatment time for the oriented reference sample; e) amorphous component orientation ($f_{c,a}$) as a function of heat treatment time for the oriented reference sample

resulted in a major decrease in orientation to *f*-values of 0.06 (crystalline) and 0.07 (amorphous).

The size of the crystallites was estimated by applying the Scherrer equation:

$$L_{hkl} = \frac{K\lambda}{\beta_0 \cos\theta}\,, \tag{4}$$

where L_{hkl} is the size of the crystal perpendicular to the (*hkl*) planes, *K* is a constant set equal to 1, λ is the wavelength of the x-radiation, β_0 is the breadth in radians at the half-maximum intensity of the pure reflection profile, and θ is the diffraction angle. The size of crystals along the normal of the hexagonal (100) [6], d-spacing 0.45 nm, for the studied samples was 11 ± 1 nm, which is in accordance with earlier data of Blundell [7].

Conclusion

This paper is the first to describe reorientation (randomization) of a semicrystalline, nematic polymer. It is demonstrated that the healing of knit lines in semicrystalline Vectra requires the melting of crystallites. The major increase in knit-line strength (healing) occurs in the melting range, the latter being confirmed by DSC and TMA. SEM and WAXS confirmed that healing was associated with reorientation (randomization) of the oriented structure. It was observed that within the melting range, randomization occurred instantaneously, accompanying a temperature increase, and that chain orientation remained unchanged on isothermal treatment. Melting of the approximately 11 nm crystallites is thus sudden and the surrounding material including the molten material randomizes rapidly. The small crystallites present in the nematic structure act as physical crosslinks which preserve the oriented structure. The fully nematic melt randomizes slowly on annealing to finally approach a globally isotropic structure.

Acknowledgements

The reported study has been sponsored by the National Swedish Board for Technical Development (STU), grant no. 87—00959, the Swedish Research Council for Engineering Sciences (TFR), grant no. 91—245, the Swedish Natural Research Council (NFR) — grant no. K-KU 1910-303, and by the Royal Institute of Technology (LUFT). The injection-molding performed at Ericsson Telecom AB, Kristianstad, Sweden is gratefully acknowledged.

References

1. Engberg K, Knutsson A, Werner P-E, Gedde UW (1990) Pol Eng Sci 30:1620—1627
2. Kausch H-H (1987) in Polymer Fracture. 2nd rev. ed., Springer-Verlag, Berlin, Heidelberg 10:416
3. Gutierrez GA, Chivers RA, Blackwell J, Stamatoff JB, Joon H (1983) Polymer 24:937—942
4. Hermans PH (1946) in Physics of Cellulose Fibres, Elsevier, Amsterdam
5. Windle AH (1982) in Developments in Oriented Polymers — 1, Applied Science, Essex 1:1
6. Chivers RA, Blackwell J (1985) Polymer 26:997—1002
7. Blundell DJ (1982) Polymer 23:359—364

Authors' address:

Prof. U. W. Gedde
Dept. Polymer Technology
Royal Institute of Technology
S-10044 Stockholm, Sweden

Effect of pressure on thermodynamic and viscosity of PS and two LC polymers

P. Izu, Mª E. Muñoz, J. J. Peña*), and A. Santamaría

Departamento de Ciencia y Tecnología de Polímeros
*) Departamento de Física de materiales, Facultad de Químicas de San Sebastian, Universidad del País Vasco, Spain

Abstract: Slit-die measurements were carried out on two commercial thermotropic LCPs (a copolyester and a copolyesteramide). Polystyrene was also tested in order to compare to a conventional polymer. The pressure profiles during the slit flow have concave and convex curvatures, depending on the polymer sample and temperature. From these curvatures the pressure coefficient of viscosity (α) was evaluated. In the case of the polyesteramide the convex curvature leads to a negative coefficient (the viscosity decreases with pressure, instead of increasing). The pressure coefficient of viscosity is also evaluated through PVT measurements using the theory of Miller.

Key words: slit-die rheometry; pressure coefficient of viscosity; liquid crystal polymer; PVT (pressure-volume-temperature) measurements

Experimental

Two ligand crystalline polymers, purchased from Hoechst Celanese, were used in this study. A copolyester of 73 mole % p-hydroxybenzoic acid and 27 mole % of 2-hydroxy-6-naphthoic acid, known comercially as Vectra A950. The other, Vectra B950, was a copolyesteramide synthesized from 2-hydroxy-6-naphthoic acid (60%), 4-aminophenol (20%), and terephthalic acid (20%). Properties and chemistry of these types of polymers have been reviewed by Calundann and Jaffe [1]. Polystyrene (Arralane, Spain, Mw = 272 000, Mn = 98 000) was also used, in order to compare the results of the rather peculiar thermotropics with a conventional polymer.

Before testing, the LCPs were dried in an oven at 100 °C for at least 24 h.

The rheological measurements were carried out using the slit-flow mode of the Göttfert Rheograph 2002 rheometer. The slit-die's dimensions were: l = 100 mm, w = 10 mm, and h = 1 mm. Three melt pressure transducers (P_1, P_2, and P_3) located, respectively, at 30 mm, 55 mm, and 80 mm from the entrance, were used to obtain the pressure profiles. The PVT operation mode of the Rheograph 2002 allows one to measure the specific volume at different constant temperatures as a function of pressure, and it gives the value of the compressibility (β). To carry out these test the die is closed by means of a toggle type lock.

Results and discussions

Pressure profiles during slit-die flow, as a function of axial distance from the exit, were obtained from the pressure transducers described in the experimental part. Figures 1, 2 and 3 show these profiles for PS, Vectra A950, and Vectra B950, respectively. Experimental data could be adjusted with a parabolic function (concave and convex). Though linear fits are very common, parabolic concave curvatures have also been reported [2—6]. That could be caused by using too shorts slits where one pressure transducer would be located very close to the entrance (the flow is not fully developed). In our case, the first transducer's location is 30 mm, where one considers that fully developed flow is achieved. Viscous heating effects are negligible, as the stress increases continuously (no maximum in the stress-time curve) up to the steady-state value

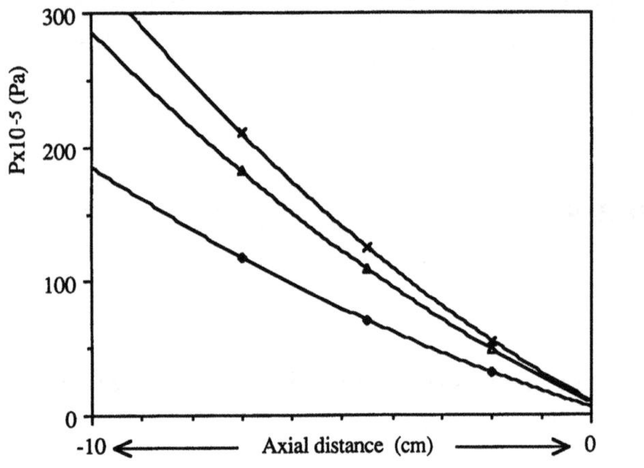

Fig. 1. Pressure as a function of axial distance for PS at 180 °C. Shear rates from top to bottom are: 285.1 s^{-1}; 152.6 s^{-1}; 34.9 s^{-1}

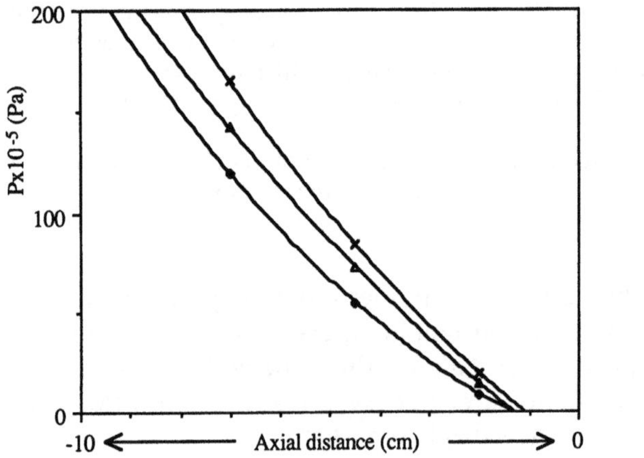

Fig. 2. Pressure as a function of axial distance for Vectra A950 at 260 °C. Shear rates from top to bottom are: 402.8 s^{-1}; 56 s^{-1}; 21.6 s^{-1}

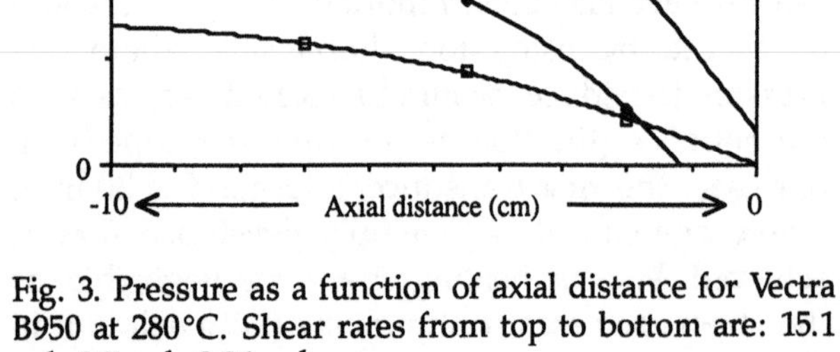

Fig. 3. Pressure as a function of axial distance for Vectra B950 at 280 °C. Shear rates from top to bottom are: 15.1 s^{-1}; 2.7 s^{-1}; 0.26 s^{-1}

(see Fig. 3.3 of [7]). The non-linearity of the pressure profiles could also be caused by pressure effects upon viscosity and volumetric flow.

We analze curvatures of the pressure profiles in terms of compressibility and dependence on pressure of the viscosity. The pressure coefficient of viscosity a describes the dependence of the viscosity upon pressure:

$$a = (\partial \ln \eta / \partial p) \tag{1}$$

With a similar development as that of Rauwendaal and Fernandez [4], and assuming that the temperature effect is negligible, the influence of the pressure upon the pressure gradient $\partial p/\partial z$ can be expressed.

$$\frac{\partial p}{\partial z} = \{1 + (a - n\beta)P\}\left\{\left(\frac{2m}{h}\right) \times \left(\frac{2MV_0(1 + 2n)}{nWh^2}\right)^n\right\}, \tag{2}$$

where β is the compressibility, n and m are the corresponding parameters of the power law model $\sigma_{21} = m\gamma^n$, M the mass flow rate, and V_0 the specific volume at $p = 0$. Integrating Eq. (1) with the condition: $z = 0$ $p = p_{ex}$ (pressure at the exit) gives:

$$1 + (a - n\beta)P = \{1 + (a - n\beta)P_{ex}\} \times \exp\{F_1(a - n\beta)z\} \tag{3}$$

where

$$F_1 = \frac{2m_r}{h}\left[\frac{2M\hat{V}_0(1 + 2n)}{nwh^2}\right]^n . \tag{4}$$

Using a polinomial series for the exponential function up to the second power allows us to express the pressure as quadratic function of z:

$$P = P_{ex} + \{1 + (a - n\beta)P_{ex}\}F_1 z + \frac{1}{2} \times \{1 + (a - n\beta)P_{ex}\}F_1^2(a - n\beta)z^2 = d + ez + fz^2 . \tag{5}$$

The pressure coefficient of viscosity is obtained from the polinomial coefficients:

$$a = n\beta + \frac{2f}{e^2} . \tag{6}$$

The compressibility (β) was obtained from PVT measurements. The apparatus gives values at dif-

Table 1. Coefficients of parabolic fits and pressure coefficient of viscosity at different shear rates for PS and Vectra A950

Polymer	γ (s^{-1})	$e \times 10^{-5}$ ($Pa \cdot cm^{-1}$)	$f \times 10^{-10}$ ($Pa \cdot cm^{-2}$)	a (Eq. (6)) (Pa^{-1})
Polystyrene	2.4	3.72	0.27	3.95×10^{-3}
	34.9	11.49	0.61	9.30×10^{-2}
	152.6	17.50	1.02	6.60×10^{-2}
	285.1	19.67	1.27	6.50×10^{-2}
Vectra A950	21.6	8.58	1.51	4.11×10^{-2}
	56.0	18.01	0.85	5.20×10^{-3}
	402.8	17.31	1.29	8.30×10^{-3}

ferent pressures. In any case, the order of the second term of Eq. (6) is higher than the order of the first (10^{-5} vs. 10^{-10}). Thus, the first term, which corresponds to the influence of pressure upon volumetric flow, can be neglected.

Table 1 summarizes the pressure coefficients of viscosity obtained for PS and Vectra A950 at different shear rates. Whereas in both cases the curvatures were concave, for Vectra B950 the convex curvature leads to negatives values of a. This striking result is in contradiction with the free volume theory.

We have also evaluated the pressure coefficient of viscosity using Miller's theory [8]. Miller thermodynamic developed the expression of a (Eq. (1)) and obtained:

$$a = \left(\frac{\partial \ln \eta}{\partial p}\right) = \left(\frac{\partial \ln \eta}{\partial T}\right)_P \times \frac{\beta V}{(\beta V)_g} \times \frac{dT_g}{dp}, \qquad (7)$$

where V is the specific volume and the subscript "g" refers to the value at the glass transition temperature.

Assuming an Arrhenius variation of the viscosity with temperature, the previous equation can be expressed:

$$a = \frac{E_a}{RT^2} \times \frac{\beta V}{(\beta V)_g} \times \frac{dT_g}{dp}. \qquad (8)$$

E_a was calculated from the flow curves obtained at different temperatures in the slit-die rheometer.

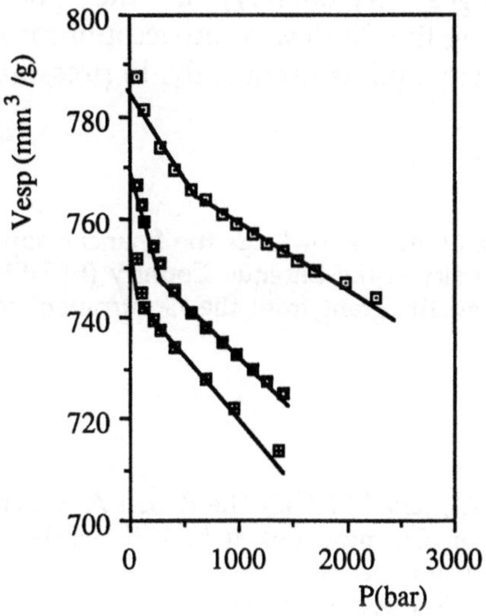

Fig. 4. Specific volume vs. pressure for Vectra A950 at three different temperatures: ⊡ at 100 °C; □ at 140 °C; ⊞ at 260 °C

The products "βV" and the variation of the T_g with pressure were obtained from PVT measurements. In an isothermal measurement there is a value of pressure where the slope of the plot specific volume vs. pressure changes (Fig. 4). Thus, we obtained a pair of values (T_g, p); by plotting these points the third term of Eq. (8) was obtained. Table 2 shows these values for PS and Vectra A950, and the values of a at higher shear rates obtained from the pressure profiles. They are of the same order as literature values [9, 10]. In the case of Vectra B950, the negatives values of a (convex curvatures of pressure profiles) could be explained by a positive variation of viscosity with tempera-

Table 2. a calculated from experimental pressure profiles and Miller's equation

Polymer	a (bar^{-1}) (Eq. (6))	a (bar^{-1}) (Eq. (7))	$\partial \ln \eta / \partial T (K^{-1})$ $(-E_a/RT^2)$	dT_g/dp (K/bar)	$(\beta V)/(\beta V)_g$
Polystyrene	6.6×10^{-3}	3×10^{-3}	6.1×10^{-2}	0.036	1.4
Vectra A950	8.2×10^{-3}	5×10^{-3}	1.11×10^{-1}	0.026	1.6

ture (in Eq. (7)) instead of the usually negative variation. However, this hypothesis is not likely unless a transition from nematic to isotropic state takes place. Our current point of view is that some structural changes (crystallinity) in the polyesteramide during the slit flow could account for a change in viscosity and, consequently, in pressure profiles.

Acknowledgements

The authors gratefully acknowledge the financial support of the University of the Baseque Country (PI UPV 203215E170/90) and the grant from the Government of Navarre to P. Izu.

References

1. Calundann GW, Jaffe M (1982) The Robert A. Welch Conferences on Chemical Research, XVI Synthetic Polymers, Houston, Texas
2. Leblanc JL (1976) Polymer 17:235
3. Laun HM (1983) Rheol Acta 22:171
4. Rauwendaal C, Fernandez F (1985) Polym Engng Sci 25:765
5. Tuna NY, Finlayson BA (1988) J Rheol 32(3):285
6. Han CD (1988) In: Collyer AA, Clegg DN (eds) Elsevier Applied Science, Rheological Measurements, London
7. Dealy JM (1982) Rheometers for molten plastics, Von Nostrand Reinhold Company, New York
8. Miller AA (1971) Macromolecules 4:757
9. Goldblatt PH, Porter RS (1976) J Appl Polym Sci 20:1199
10. Tuna NY, Fynlanson BA (1988) J Rheol 32(3):285

Authors' address:

Anton Santamaría
Departamento de Ciencia y Tecnología de Polimeros
Facultad de Química
Apartado de correos 1072
E-20080 San Sebastian (GUIPUZCOA), España

Amorphous polymers

Interpretation of structural recovery of amorphous polymers from DSC data

J. M. Hutchinson

Department of Engineering, Aberdeen University

Abstract: The response of amorphous polymers on heating at constant rate in the differential scanning calorimeter is analysed as a function of the whole thermal history of the polymer. First, it is shown that a quantitative analysis of the data should make allowance for thermal gradients in the sample, which tend to broaden the response. Subsequently, the data are interpreted on the basis of the Tool-Narayanaswamy approach and involving three material parameters: an activation energy, Δh^*; a parameter x ($0 < x \leqslant 1$), determining the relative contributions of temperature and structure to the retardation times; and an exponent β ($0 < \beta \leqslant 1$) of the stretched exponential response function, which is inversely proportional to the width of the distribution of retardation times. The activation energy can be found from the dependence of the glass transition temperature on cooling rate, or from the shift of the endothermic peak temperature with heating rate in intrinsic thermal cycles with fixed ratio of cooling to heating rate. The latter cycles provide also a means of correcting the data for thermal lag in the sample. The parameter x can be found from the dependence of the endothermic peak temperature on either annealing time or heating rate, provided that the glass is well annealed prior to scanning. On the other hand, if the glass is only poorly annealed, a quite different endothermic peak is observed, the height of which is sensitive to the parameter β. These material parameters are evaluated for polystyrene and polyvinylchloride, and are compared with those obtained for an inorganic glassy system in order to provide some insight into their physical significance.

Key words: Glasses; structural recovery; DSC; glass transition; enthalpy relaxation

Introduction

The use of differential scanning calorimetry (DSC) is widespread in the study of the thermal properties of polymers, and, in particular, in the investigation of the behaviour of amorphous polymers in the glass transition region. Most commonly, this involves heating the polymer glass through the transition region at a constant rate and observing the response. However, this heating scan is, alone, insufficient to allow a useful interpretation of the data, because the glass at the start of the heating scan is not in an equilibrium state. In other words, the response on heating depends on the state of the glass, and in a very intricate way, as will be shown below. The details of this response, however, provide a rich source of information from which a better understanding of the glass transformation process may be derived. Fundamental to this interpretation, therefore, is a knowledge of the thermal history of the polymer glass.

The usual approach is to adopt a three-step thermal cycle (Fig. 1) in which: i) the polymer is cooled at constant rate q_1 from the equilibrium melt to a temperature T_1 in the asymptotic glassy region; ii) the glass is annealed at T_1 such that the enthalpy reduces by an amount δ_H; and iii) the glass is reheated at constant rate q_2 until an equilibrium melt is again achieved. This last stage corresponds to the usual DSC heating scan, and results in the well-known endothermal peak occurring at a temperature T_p.

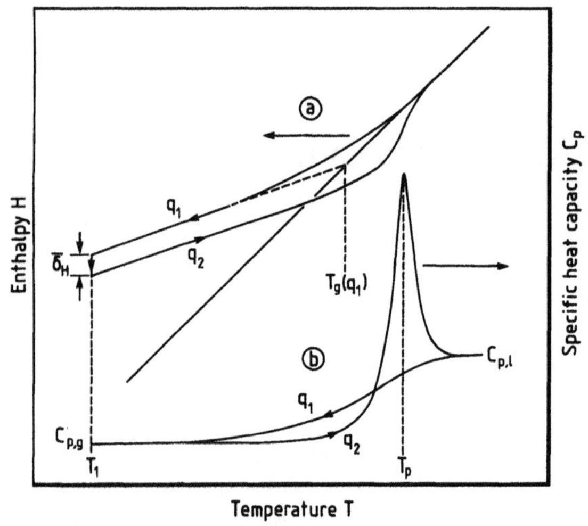

Fig. 1. Schematic illustration of the variation of enthalpy (a) and specific heat (b) with temperature in a three-step thermal cycle

The state of the annealed glass prior to reheating can be characterised by the fictive temperature, T_f, while the state of the glass formed on cooling at rate q_1 is characterised by the glass transition temperature, $T_g(q_1)$. It is clear that the peak temperature T_p is neither the fictive temperature nor the glass transition temperature; in fact, T_p is a function of the experimental variables defining the thermal cycle, namely q_1, δ_H and q_2 (T_p is independent of T_1 provided T_1 lies within the asymptotic glassy region, as is the case here). It is the dependence of T_p on these experimental variables which forms the basis for the interpretation of DSC data discussed here.

Theoretical dependence of T_p

In addition to the experimental variables, T_p will depend also upon those material parameters which control the rate of structural recovery in the glass. These parameters characterise the two important features of structural recovery [1]: a dependence of retardation times τ_i on both temperature (T) and structure (T_f), and the existence of a distribution of retardation times. Commonly, the first is expressed by an equation often attributed to Tool [2] and Narayanaswamy [3]:

$$\tau_i = A_i \exp\left[\frac{x\Delta h^*}{RT} + \frac{(1-x)\Delta h^*}{RT_f}\right], \tag{1}$$

in which x ($0 < x \leqslant 1$) determines the relative contributions of temperature and structure to the retardation times, and Δh^* is an activation energy. An equivalent equation was proposed by KAHR [4, 5]:

$$\tau_i = A_i \exp[-\theta T - (1-x)\theta\delta_H/\Delta C_p] , \tag{2}$$

in which ΔC_p is the difference between the heat capacities of the liquid and glass, and θ is related to Δh^* by:

$$\theta = \frac{\Delta h^*}{RT_g^2} . \tag{3}$$

The second feature of structural recovery can be represented either as a discrete distribution, as in the KAHR model [5], or as a continuous spectrum using the stretched exponential response function:

$$\phi(t) = \exp[-(t/\tau_0)^\beta] , \tag{4}$$

in which β ($0 < \beta \leqslant 1$) is inversely proportional to the width of the distribution and τ_0 is given by Eq. (1) or (2).

The dependence of T_p on each of the experimental variables defines a set of shifts:

$$\hat{s}(Q_1) = \theta\left(\frac{\partial T_p}{\partial \ln|q_1|}\right)_{T_1,\delta_H,q_2} \tag{5}$$

$$\hat{s}(\bar{D}) = \Delta C_p\left(\frac{\partial T_p}{\partial \delta_h}\right)_{q_1,T_1,q_2} \tag{6}$$

$$\hat{s}(Q_2) = \theta\left(\frac{\partial T_p}{\partial \ln q_2}\right)_{q_1,T_1,\delta_H} . \tag{7}$$

For well-annealed glasses, these shifts can be shown [6] to be inter-related and strong functions of x:

$$-\hat{s}(Q_1) = \hat{s}(Q_2) - 1 = \hat{s}(\bar{D}) = F(x) , \tag{8}$$

in which the caret signifies a so-called "main peak" and where the dependence of $F(x)$ on x defines the "master curve". For poorly annealed glasses, on the other hand, one finds [6]:

$$s_u(Q_1) + s_u(Q_2) = 1 ; \quad s_u(\bar{D}) = 0 , \tag{9}$$

in which the subscript u signifies an "upper peak" and where the shifts are essentially independent of x.

The difference between main and upper peaks for well and poorly annealed glasses, respectively, is crucial to a full interpretation of DSC data.

Experimental dependence of T_p

Intrinsic cycles

Intrinsic cycles are a special case of those illustrated in Fig. 1 in which $\bar{\delta}_H = 0$. Since the fictive temperature of the glass at T_1 can be found for intrinsic cycles, either by an algebraic [7] or graphical [8] method, the activation energy Δh^* is readily determined from the slope ($= -\Delta h^*/R$) of the linear relationship between $\ln|q_1|$ and $1/T_f$.

These same intrinsic cycles allow another means of evaluating Δh^* at the same time as providing a method of correcting the data for thermal lag in the sample. It can be shown theoretically [5] that intrinsic cycles for which the ratio (ρ) of cooling rate to heating rate is constant should be superposable by means of a shift along the temperature axis by an amount dependent upon θ (or Δh^*):

$$\left(\frac{\partial T_p}{\partial \ln q_2}\right)_{\rho,\bar{\delta}_H=0} = \theta^{-1} = \frac{RT_g^2}{\Delta h^*} . \quad (10)$$

In practice, though, superposition cannot exactly be achieved because the DSC response broadens at faster heating rates as a result of thermal lag in the sample [9]. The theoretical invariance of the endothermic peak shape, however, enables correction factors to be determined for each heating rate, and then θ or Δh^* can be found from the corrected peak temperature shifts. Interestingly, for some ionically conducting iodomolybdate glasses with different amounts of silver iodide, a trend of reducing thermal lag was observed as the AgI content increased, implying that the thermal conductivity was mirrored by the ionic conductivity, both increasing with AgI content [10].

Well annealed glasses

In contrast to the use of intrinsic cycles for the evaluation of Δh^*, the evaluation of x requires well annealed glasses. Under such circumstances, the evaluation of any of the shifts defined in Eqs. (5) to (7) permits the determination of x from Eq. (8) and the master curve. For example, for polystyrene (PS) the dependence of T_p (corrected for thermal lag in the sample) on $\bar{\delta}_H$ is shown in Fig. 2. It is clear that for well annealed glasses (large $\bar{\delta}_H$) there is a linear dependence of T_p on $\bar{\delta}_H$ from which $\hat{s}(\bar{D})$ and hence x may be evaluated. This approach has been applied to other polymer glasses, including polyvinylchloride (PVC), as well as to the inorganic system of iodomolybdates with 40, 50, and 60 mol % of AgI, and the values of x (and other parameters) are tabulated further below.

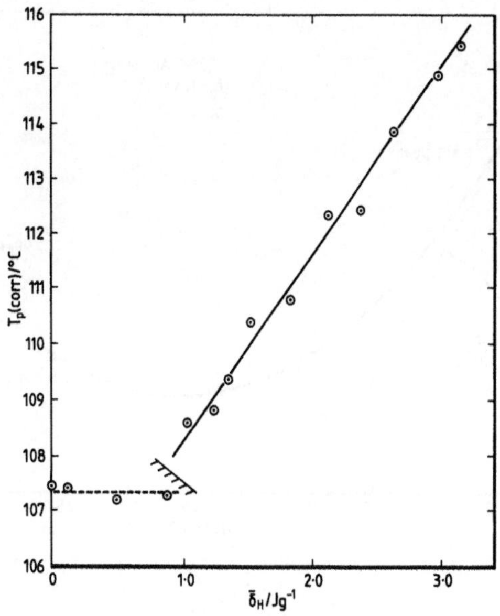

Fig. 2. Corrected peak temperature versus $\bar{\delta}_H$ for PS, obtained from a set of DSC traces with cooling rate —20 K/min, annealing temperature 85 °C, heating rate 15 K/min and annealing times up to 960 h

Poorly annealed glasses

It is also evident from Fig. 2 that the linear relationship between T_p and $\bar{\delta}_H$ does not hold for small $\bar{\delta}_H$, as indicated by the break in the line. This is because the peak in this region of small $\bar{\delta}_H$ is an upper and not a main peak, resulting from the reheating of a poorly annealed glass. Furthermore, the peak temperature is essentially independent of $\bar{\delta}_H$, as predicted by Eq. (9). This distinction between main and upper peak is also well illustrated by examining the dependence of T_p on cooling rate in intrinsic cycles: a very slow cooling rate corresponds to a well annealed glass, whereas a fast cooling rate will yield a poorly annealed glass. Figure 3 shows this for an inorganic glass: for slow cooling rates the slope is negative, from which $\hat{s}(Q_1)$ (Eq. (5)) could be obtained and hence x evaluated through Eq. (8), whereas for fast cooling rates the slope is positive and gives $s_u(Q_1)$, independent of x for these upper peaks.

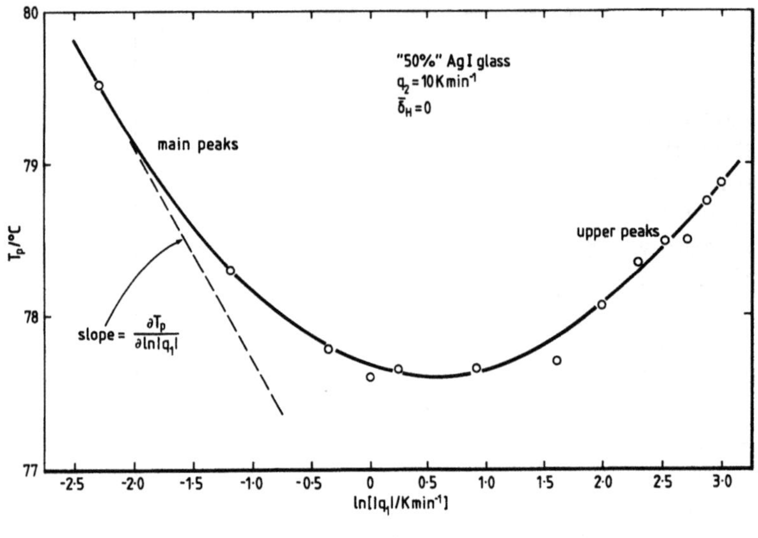

Fig. 3. Dependence of T_p on log (cooling rate) for an inorganic glass in intrinsic cycles with heating rate 10 K/min

It is interesting to examine the magnitude of the upper peaks. Defining a normalised peak height, C_{pu}^N, as the height of the peak at the upper peak temperature, T_u, divided by ΔC_p at T_u, one can obtain information about the breadth of the distribution of retardation times as follows. C_{pu}^N depends on both cooling and heating rate (more precisely, on the ratio ρ) and has a minimum value of unity. Its theoretical variation is shown in Fig. 4, for various values of β and x [11]. For all values of these parameters, the height diminishes with increasing ρ, an effect which has been observed in PS [11] and in several inorganic glasses. Furthermore, given a value of x obtained by the peak-shift method, a comparison of experimental values of C_{pu}^N with this theoretical variation will provide an estimate of β for any glass.

For PVC this approach yields some interesting results, since no upper peak is observed for $\rho \geqslant 1$ approximately; there is only a sigmoidal "step change" in C_p, as has been reported on several occasions. The implication is that β for PVC must be very small, which is indicative of a very broad distribution. This result and those for PS and the inorganic system are summarised in Table 1 together with the other parameter values.

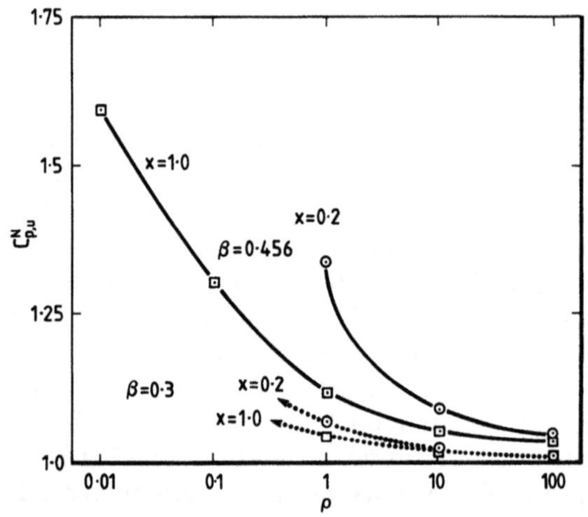

Fig. 4. Theoretical dependence of C_{pu}^N on ρ for intrinsic cycles for values of x and β indicated

Table 1. Values of the material parameters Δh^*, x, β and ΔC_p for some polymer and inorganic glasses

	PVC	PS	$(AgI)_y(Ag_2MoO_4)_{1-y}$		
			40%	50%	60%
x	0.27	0.46	0.50	0.54	0.65
Δh^*/kJ · mol^{-1}	1100	580	640	505	360
ΔC_p/Jg^{-1} K^{-1}	0.335	0.281	0.164	0.139	0.099
β	<0.3	0.456 < β < 0.6	>0.456 — increasing with y		

Results and discussion

In order to attempt to interpret the meaning of these parameter values, particularly of x and β, one can seek some correlations between them. For both polymers and for the inorganic glasses, higher values of x are associated with lower values of Δh^*, a correlation that has been observed before [12]. On the other hand, they also correlate with lower values of ΔC_p, implying that for increasing x there is less of a difference between melt and glass structure. This has led to the term "continuity parameter" for x, measuring the degree to which the melt structure is retained in the glass [10]. This is consistent with a cluster-tissue model for inorganic glass [13], in which increasing AgI content leads to increasing amounts of tissue (or melt-like structure) in the glass; at the same time, the ionic conductivity of the glass also increases, which is explained on the basis of transport of silver ions in the tissue of the glass.

Similar correlations can be seen for the parameter β. In particular, β decreases with increasing Δh^* in both the polymer and inorganic glasses. This inverse correlation, which has been noted before [12], suggests that low values of β, and hence broad distributions, result from an increased degree of co-operativity of molecular motion perhaps involving a larger number of chain segments. The larger values of β for the inorganic glasses is consistent with the idea that structural recovery in these materials is less co-operative than in polymers [12].

Conclusions

These results show that the C_p traces obtained by DSC on amorphous polymers (or, indeed, any glasses) are a rich source of information for the interpretation of structural recovery. However, some care must be exercised in their analysis, for a number of reasons. First, a quantitative analysis must allow for the effects of thermal lag in the sample. Second, different experimental conditions can give rise to very different responses, which reflect quite different material characteristics. In particular, will annealed glasses give C_p peaks which occur at temperatures dependent primarily upon the parameter x, which controls the influence of structure on the retardation times, whereas poorly annealed glasses give C_p peaks whose magnitude is strongly dependent upon the parameter β, which provides a measure of the breadth of the distribution of retardation times. The evaluation and interpretation of these material parameters for a wide range of glasses, both polymeric and inorganic, is considered to be a useful approach towards a better understanding of the amorphous state.

References

1. Kovacs AJ (1963) Fortschr Hochpolym Forsch 3:394–508
2. Tool AQ (1945) J Am Ceram Soc 29:240–253
3. Narayanaswamy OS (1971) J Am Ceram Soc 54:491–498
4. Kovacs AJ, Hutchinson JM, Aklonis JJ (1977) In: Gaskell PH (ed) The Structure of Non-Crystalline Materials. Taylor and Francis, London, pp 153–163
5. Kovacs AJ, Aklonis JJ, Hutchinson JM, Ramos AR (1979) J Polym Sci Polym Phys Ed 17:1097–1162
6. Ramos AR, Hutchinson JM, Kovacs AJ (1984) J Polym Sci Polym Phys Ed 22:1655–1695
7. Richardson MJ, Savill NG (1975) Polymer 16:753–757
8. Moynihan CT, Easteal AJ, De Bolt MA, Tucker J (1976) J Am Ceram Soc 59:12–16
9. Hutchinson JM, Ruddy M, Wilson MR (1988) Polymer 29:152–159
10. Pappin AJ, Hutchinson JM, Ingram MD (1991) Phys Chem Glasses 32:121–128
11. Hutchinson JM, Ruddy M (1990) J Polym Sci Polym Phys Ed 28:2127–2163
12. Hodge IM (1983) Macromolecules 16:898–902
13. Ingram MD (1989) Phil Mag B 60:729–740

Author's address:

Dr. J. M. Hutchinson
Department of Engineering
Aberdeen University
Fraser Noble Building
King's College
Aberdeen AB9 2UE, U.K.

Enthalpy relaxation in amorphous and crystalline polymers. Polyaryl ether ether ketone blends

J. N. Hay

The School of Chemistry, The University of Birmingham, U.K.

Abstract: The kinetics of enthalpic relaxation in compatible blends of polyether ether ketone, PEEK and polyetherimide, PEI have been measured by DSC, and interpreted in terms of a broad spectrum of relaxation processes, in that the overall fractional extent of conversion $\phi(t)$ at time t obeys the relationship,

$$1 - \phi(t) = \exp(-t/\tau)^{\beta} ,$$

where τ is the relaxation time and β is a measure of the reciprocal breadth of the relaxation spectrum. Similar analysis has been carried on the dynamic and dielectric response of the blends, but there is little similarity in the β values so determined. Although the 50:50 blend exhibits a single glass transition temperature, the observed value of β, 0.2 for the dielectric relaxation spectra is consistent with some heterogeneity developing at the molecular level.

Key words: Physical aging; DSC; dielectric relaxation; polyetherimide; PEEK; blends

Introduction

Organic glasses are not normally in thermal equilibrium but approach it over extended periods in a temperature region below but close to the glass transition temperature, T_g. Physical aging is attributed to the progressive change in the glass towards equilibrium, the restricted mobility of chain segments as the temperature is reduced below the transition, and is associated with the formation of the glass from the liquid on cooling.

Since the material properties of the quenched glass change progressively the process has considerable commercially importance and has been widely studied. Conventionally it has been observed that there is an increase in the yield stress and tensile modulus, and a decrease in enthalpy, specific volume and fracture toughness of the glass with increasing extent of physical aging. Changes in material characteristics, such as specific volume, enthalpy and creep compliance, are conveniently used to follow the process of physical aging, but exact correlation between the experimental procedures have not been observed. This was stressed recently [1] when the difference in the temperature range over which physical aging can be detected by creep measurements, over 250 K, and by changes is enthalpy, about 50 K were discussed. under these circumstances, it is convenient to use the term enthalpic relaxation and restrict it to calorimetric studies of physical aging. Differential scanning calorimetry appears to be best suited for such studies.

We have used enthalpy relaxation to probe the microtexture of compatible blends, and the present report investigates the miscibility of polyether ether ketone/polyetherimide compatible blends as a function of composition.

Experimental

A commercial grade of PEI was used, Ultem 1000, manufactured by GEC. The polymer has a number average molecular weight of about 20 000. It was dried at 150°C for 4 h prior to use. A commercial grade of PEEK was also used, as supplied by ICI Polymer and Chemicals Ltd. It was used and characterised as described elsewhere [2]. Blends were produced by weight, on dried polymer, using an APV model MP2000 twinscrew extruder at 380°C. Sheets, 1 × 150 × 150 mm, of amorphous blends were moulded at 350°C using a hydraulic press, dried, and quenched directly into ice/water. Required samples were cut directly from these sheets.

Thermal analysis was carried out on a Perkin-Elmer differential scanning calorimeter, model DSC-2C, interfaced to and controlled by a BBC Master computer. Details of the procedures involved in measuring the glass transition temperature and following the kinetics of physical ageing have been discussed elsewhere [1].

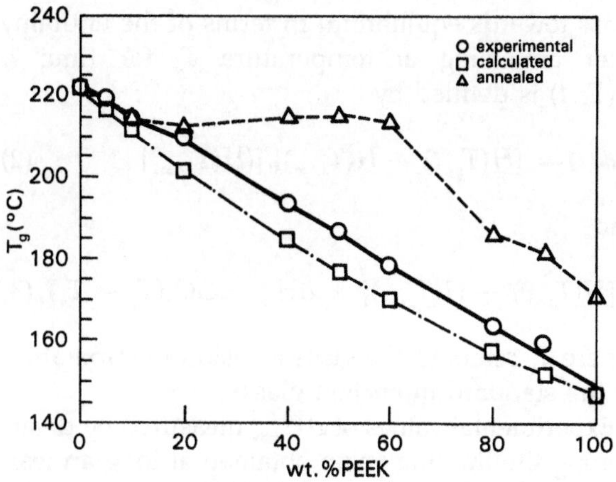

Fig. 1. The variation of glass transition temperature with composition for PEEK/PEI blends. Amorphous samples crystallised at temperature for which the crystallisation half-life was 30 min. Measured by DMTA at 3 K min^{-1} and at 1 Hz. Calculated from the Fox equation

Results and Discussion

The blends were initially examined by DMTA between 20° and 250°C at a range of frequencies between 1—10 Hz, and a heating rate of 3 Kmin^{-1}. The peak temperature in tanδ at 1 Hz was taken to define the T_g and this changed progressively with composition between the homopolymer values (see Fig. 1). Crystallization of PEEK from the blend occurred immediately above T_g with a corresponding increase in modulus and tanδ values. Crystallisation increased the T_g due to the amorphous phase containing less PEEK and the reinforcing effect of the PEEK crystals. The values of the T_g of the amorphous samples were substantially different from those calculated by the Fox relationship.

$$(T_g)^{-1} = x_1/T_{g1} + (1 - x_1)/T_{g2} \, , \qquad (1)$$

where x_1 is the weight fraction of homopolymer 1, and T_{g1} and T_{g2} the glass transition temperature of the two homopolymers 1 and 2 respectively. The negative deviation is consistent with specific interactions between the components, and the presence of a single T_g is consistent with miscibility of the two homopolymers over the total composition range.

A single T_g was observed for all blend compositions by DSC and the progress of enthalpic relaxation was monitored by the development of the endothermic peak (see Fig. 2). The relaxation of the

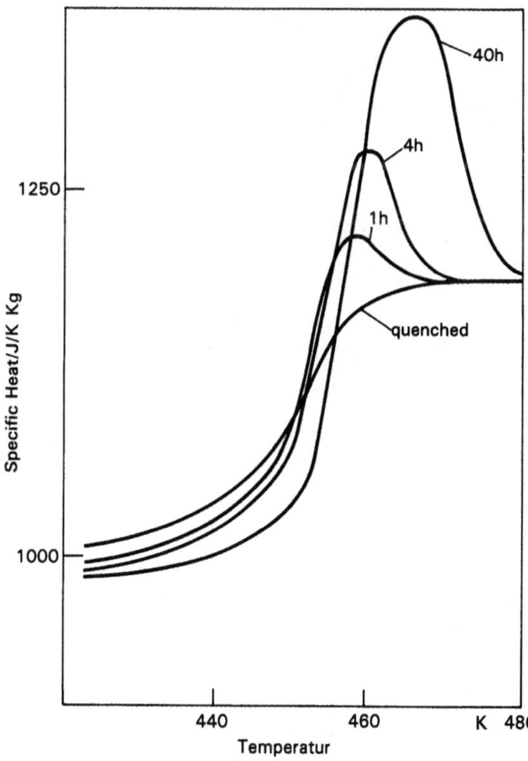

Fig. 2. The change in heat capacity of 50% blend on aging for various times at 440 K

glass towards equilibrium in terms of the enthalpy after annealing at temperature T_a for time t, $H(T_a, t)$ is defined by

$$\phi(t) = [H(T_a, t) - H(T_{a,\infty})]/[(H(T_{a,\infty})] \quad (2)$$

and

$$[H(T_a, 0) - H(T_{a,\infty})] = \Delta H_\infty = \Delta C_P(T_g - T_a), \quad (3)$$

where T_g refers to the glass transition temperature of the standard quenched glass.

Experimental values of ΔH_{max} measured by determining the limiting value obtained at long annealing times, at low ΔT up to 15 K (Eq. (3)) were accurately followed. Indeed, plots of ΔH_{max} against T_a were linear and could be used to determine the value of the T_{ag}(quenched), which compared favourably with the limiting value of ΔH_t determined at excessively long aging times. At high supercooling ΔH_{max} did not follow the relationship and appeared to be decreasing markedly with increasing ΔT (see Fig. 3).

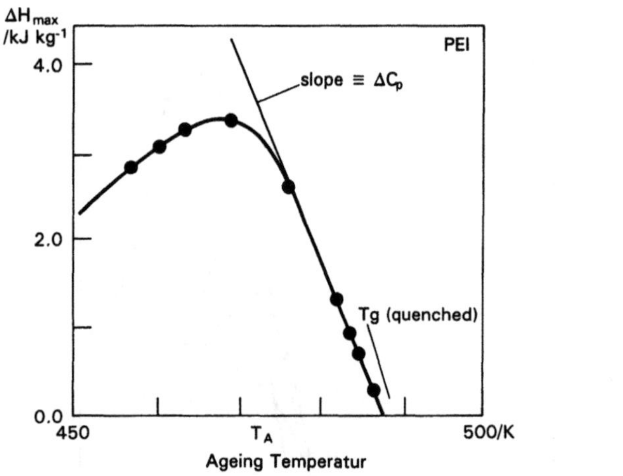

Fig. 3. The variation of ΔH_{max} determined experimentally with aging temperature, T_a in PEI

The temperature range over which enthalpic relaxation occurred was found to be limited to T_g and $T_g - 50$, in that, within experimental error, no endothermic peaks could be detected on aging over extended periods at ΔT values of 60 and 70 K.

The extent of enthalpy relaxation increased logarithmically with time, and was markedly dependent on temperature. The overall extent of enthalpy relaxation with time was analysed in terms

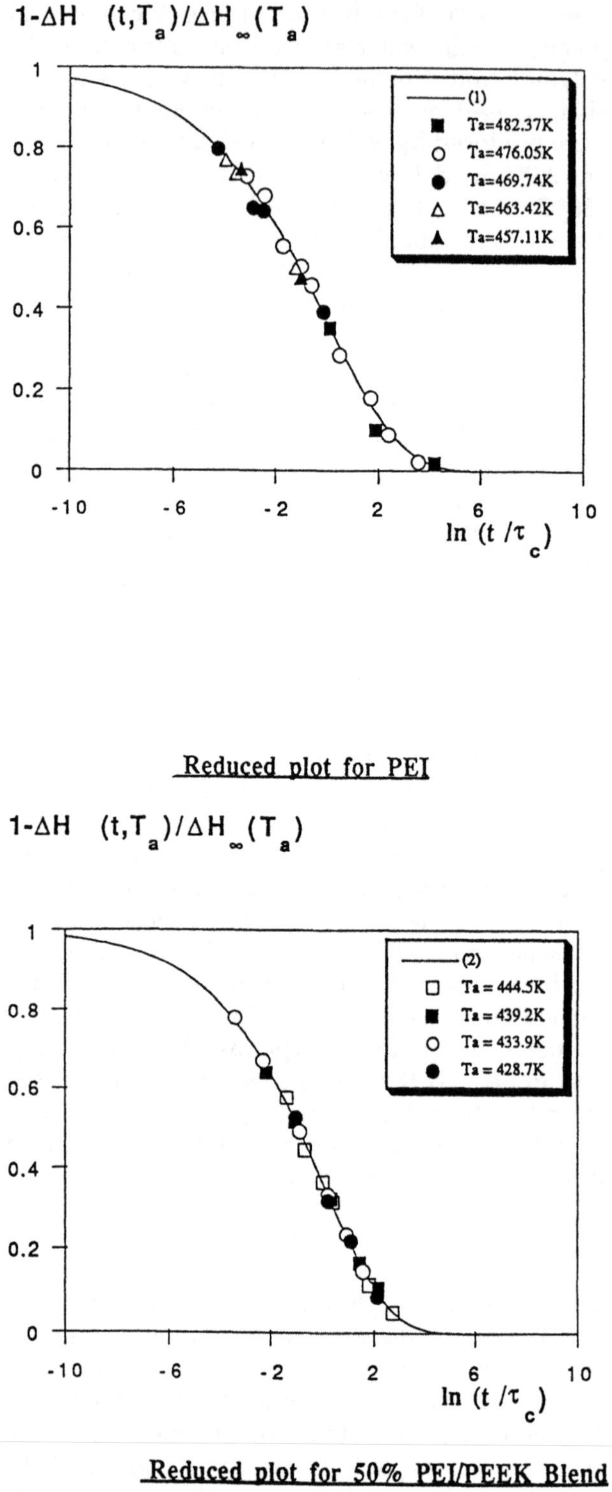

The full lines represent the theoretical curves with (1) β = 0.35

(2) β = 0.40.

Fig. 4. Reduced plot of $\phi(t)$ against $\ln(t/\tau)$. 1) PEI $\beta = 0.35$; 2) 50% blend, $\beta = 0.40$

of Moynihan's approach [3] in which $\phi(t)$ was given by the William-Watts form [4],

$$\phi(t) = \exp\{-t/\tau\}^{\beta} , \quad (4)$$

where $\beta(0 < \beta < 1)$ was inversely related to the breadth of the distribution and τ is the characteristic retardation time. For each blend the enthalpy relaxation data could be fitted to a master curve of $\phi(t)$ against $\ln(t/\tau)$ (see Fig. 4) and an average β value could be determined for each blend from the master curve (see Table 1).

Table 1. β values

Blend % PEEK	Temperature range K	β value
a) Enthalpy relaxation		
100	389—410	0.55—0.6
50	429—445	0.40
0	457—482	0.35
b) Dielectric relaxation		
100	421—424	0.33
50	400—469	0.13—0.22†
50	493—505	0.12—0.18*
00	503—523	0.36 ± 0.03
c) DMTA		
100	413—428	0.47 ± 0.03†
100	446	0.11*
50	457—461	0.40 ± 0.06
0	493	0.53

* Partially crystalline.
† Amorphous.

Variations were observed depending on the value of ΔH_{max} chosen, but the β values for physical aging varied between the limits of 0.35 and 0.6 set by the two homopolymers and they changed progressively with composition.

In analysing the dielectric response of polymers the William-Watts relationship has been widely used. The dielectric relaxation behaviours of the homopolymers and blends were measured by DETA as a function of frequency at constant temperature. Following the treatment of Havrillak [5], corresponding to the superposition of Debye-like relaxation processes with a range of relaxation time,

$$\varepsilon^* = \varepsilon + [(\varepsilon_0 - \varepsilon)/(1 + i\omega t)^{\beta}] , \quad (5)$$

where ε^* is the measured permittivity at the angular frequency ω, ε_0 is the static permittivity, and ε the high frequency permittivity. Cole-Cole plots were used to determine the dielectric β value from a curve-fitting procedure. The parameters are listed in Table 1.

β values were also calculated from a similar analysis of DMTA data, but the later values determined at lower frequency were consistently higher than the equivalent DETA values, and similar to the β values determined by analysis of the enthalpy relaxation behaviour. Extremely low β values were determined for the dielectric response of the blends, 0.1—0.2, which implied that the dipole relaxation spectra has markedly broadened in the blends. The extent of the broadening is consistent with a heterogeneous system, and a composite blend from which PEEK has crystallised exhibits similar low β values.

Conclusions

The dielectric relaxation spectra of compatible blends broaden with composition consistent with the developed of some heterogeneity at the molecular level. Enthalpic relaxation does not exhibit this effect, and the broadening of the relaxation spectra is consistent with the change in composition between the two homopolymers.

References

1. Hay JN, Kemmish DJ (1985) Polymer 26:905
2. Goodwin AA, Hay JN (1990) Polymer Comm 31:337
3. Moynihan CT (1976) Ann N.Y. Acad Sci 279:15
4. Williams G, Watts DC (1970) Trans Faraday Soc 66:80
5. Havrillak SJ (1966) J Polymer Sci 14:99

Author's address:

J. N. Hay
School of Chemistry
University of Birmingham
Birmingham B15 2TT, U.K.

Structural relaxation process on a thermosetting resin*)

S. Montserrat Ribas

Laboratori de Termodinàmica i Físico-química,
E.T.S. Enginyers Industrials de Terrassa, Universitat Politècnica de Catalunya, Spain

Abstract: Structural relaxation process in a fully cured epoxy resin is investigated by differential scanning calorimetry (DSC). Samples with a glass transition temperature of 98.5°C ($T_{g(1/2)}$) are submitted to thermal cycles involving a quenching through glass transition region, an isothermal aging at a temperature ($T_a < T_g$) for various times (t_a), and a reheating at 10 K/min to a temperature above T_g. When T_a is well below T_g (T_a = 50 to 70°C) a sub-T_g peak appears whose size and position are strongly dependent on aging conditions (T_a and t_a). A peak superposed to T_g appears when the samples are aged at T_a somewhat lower than T_g (T_a = 80 and 90°C). In this case, the position of the peak is less dependent of t_a. The behavior of these endothermic peaks must be attributed to the kinetics of the structural relaxation process, and not to some change of the morphology of the crosslinking of polymer network.

Key words: Structural relaxation; physical aging; thermosets; epoxy resin; DSC method

Introduction

Amorphous polymers cooled to below their glass transition temperature (T_g) undergo a structural relaxation process in which their properties, such as volume, change with time. This process, first studied by Kovacs [1] on polyvinyl acetate by volume relaxation, is a consequence of the non-equilibrium nature of the glassy state, which evolves toward a metastable equilibrium state at a rate that depends on the temperature, pressure, and thermal history of the glass. More recently, this process has been called physical aging by Struik [2]. A decrease in enthalpy of the sample has been detected in line with this decrease of volume.

It is well known that the extension of the structural relaxation process can be monitored by differential scanning calorimetry (DSC). An endothermic peak is detected in the DSC curve (heat capacity or heat flow measurement) when a sample which has been aged isothermally at a temperature T_a (usually called the aging temperature), lower than the T_g, is reheated through the glass transition region. It has been shown by the DSC technique that the temperature, size, and intensity of this endothermic peak is dependent on the previous thermal history of the sample.

Since the work of Kovacs, the structural relaxation process has been studied extensively in linear amorphous polymers, but less so in thermosetting polymers. Kong [3] thinks that this is due to the difficulty of carefully controlling the crosslink density of the network and the precautions needed to avoid chemical aging of the resin.

Structural relaxation studies on fully cured epoxy resins were reported in an earlier paper by Ophir et al. [4], and more recently by Kwong et al. [5] and Plazek et al. [6]. These works show that thermosetting materials can undergo structural relaxation in a similar way to thermoplastics.

In this work, results of the structural relaxation process of a fully anhydride cured epoxy resin monitored by differential scanning calorimetry (DSC) are reported. The isothermal aging tempera-

*) Presented at 23rd Europhysics Conference on Macromolecular Physics "Solidification Processes in Polymers", Stockholm, 1991.

ture range was from T_g-50 to a temperature very near to T_g, with periods of time from 15 min to 70 days, depending on the aging temperature. The calorimetric results show the structural relaxation of the sample through the behavior of the endothermic peak which appears as a consequence of the enthalpy recovery process.

Experimental part

Materials and curing procedure

The epoxy resin was a modified diglycidyl ether of bisphenol A (CIBA-GEIGY Araldite CY225) with an epoxy equivalent of 188 g/equiv. A hardener derived from phthalic anhydride and with an accelerator (CIBA-GEIGY HY225) was used to cure the resin. Both components were used as received and mixed at a weight ratio of 10:8 (resin:hardener). The mixture was stirred at room temperature for 20 min, and then degassed in a vacuum chamber for 15 min. About 10 mg of sample were enclosed in aluminum DSC pans, placed in Pyrex tubes under nitrogen atmosphere, and thermostated at 130°C for 8 h. These curing conditions ensure that the resin becomes practically fully cured, such that no residual exothermic reaction was detected by heating the cured sample up to 260°C. The kinetics of the curing process have been reported in a previous paper [7].

Differential scanning calorimetry

A Mettler Thermoanalyser TA 4000 equipped with a low-temperature-range DSC 30 Differential Scanning Calorimetry module, a TC 11 microprocessor, and the graphics software Graphware TA72 was used to measure the heat flow as a function of temperature. The calorimeter was previously calibrated with indium standards. The DSC curves were obtained using 10 mg of sample, and a heating rate of 10 K/min.

The samples were submitted to a thermal cycle of rapid cooling through the glass transition region, isothermal aging, and reheating at a slower rate. The thermal history can be summarized as follows: a) encapsulated samples were previously heated at a temperature T_0 = 130°C for 5 min, in order to erase any previous thermal history of the sample, and then quenched to the aging temperature T_a; b) aging the sample during a time (t_a) at T_a; c) quenching the sample in air down to T_1 = 20°C, transferred into the DSC, and then scanned at a heating rate of 10 K/min from T_1 to T_0. A DSC curve of output power form versus temperature was obtained.

The glass transition interval was measured in an unaged sample whose thermal history is identical to the above described, but with a null aging time (t_a = 0). The characteristic temperatures of the glass transition region in the DSC curve obtained by heating at 10 K/min are the following:

- extrapolated onset glass transition temperature, T_{g0} = 93.5°C;
- temperature of the halfpoint between the rubbery and glassy lines, $T_{g1/2}$ = 98.5°C;
- end temperature of the glass transition, T_{gend} = 103°C.

The values measured in a cooling DSC curve at 10 K/min from 130° to 20°C are very close to these: 92.3°, 98°, and 103.5°C. In this work the $T_{g1/2}$ = 98.5°C is considered as the value of the glass transition temperature of the epoxy resin.

Results and discussion

Figure 1 shows the DSC curves in the form of output power as a function of temperature for different aging temperatures from 50° to 70°C and different aging times from 1 to 264 or 1764 h depending on T_a. In some cases, these curves show the presence of two endothermic peaks. The results of aged samples at temperatures from 80° to 100°C are presented in Fig. 2. These curves only show the presence of one peak superposed on the upper part of the glass transition.

Experimental results show that the size, position, and intensity of the endothermic peak depend on the aging conditions. For T_a = 50°C, the peak appears at a temperature which depends on the annealing time t_a. For $t_a \leqslant 10$ h, the peak appears well below the glass transition interval; for t_a between 10 and 408 h, the peak has grown like a shoulder which appears between the lines corresponding to the glassy and rubbery states, respectively; and for $t_a > 408$ h the shoulder increases in magnitude and appears clearly as a peak superposed on the high part of T_g. The position and intensity of this peak are strongly dependent on t_a.

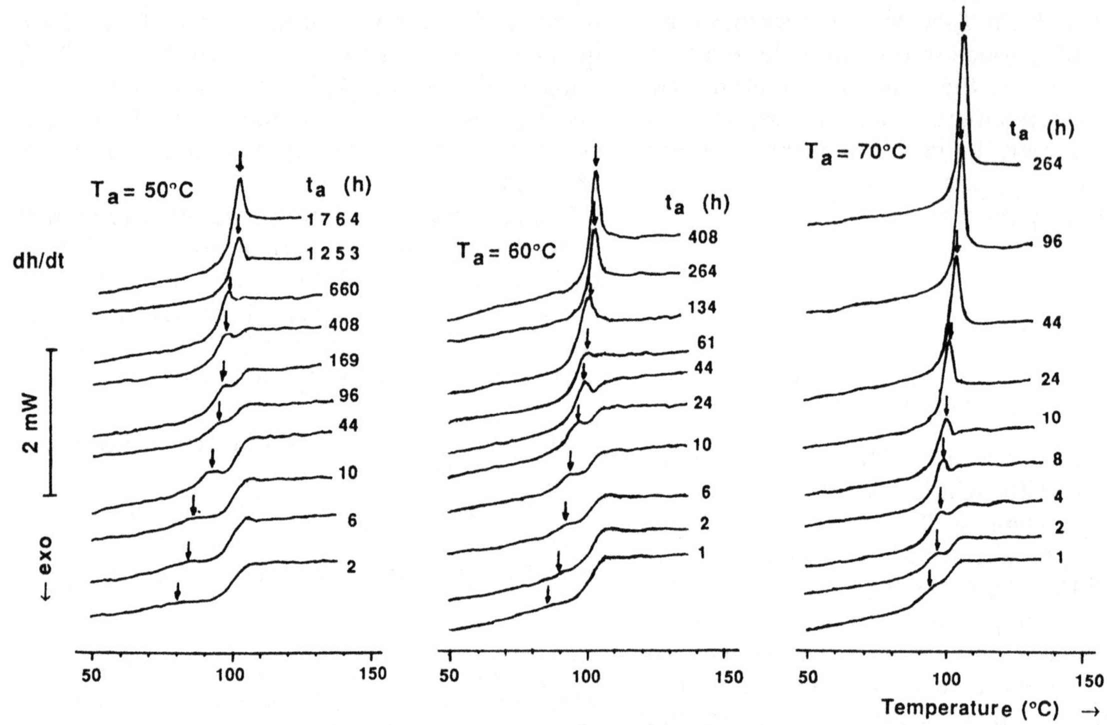

Fig. 1. DSC curves obtained at a heating rate of 10 K/min of an aged epoxy resin at temperatures between 50° and 70°C for different aging times. The arrows indicate the approximate position of the endothermic peak (or shoulders)

At T_a = 60°C a similar behavior is detected, although with differences in the aging time where the phenomenon occurs. Thus, up to t_a = 2 h the peak develops below the glass transition interval. Between 2 and 61 h, the peak appears as a shoulder in this interval, and for times higher than 61 h the peak appears superposed on the rubbery line of the glass transition. For T_a > 70°C, the peak only appears superposed on the glass transition interval. The maximum intensity of the peak appears when aging temperature is 80°C. At T_a = 90°C, the intensity of the endothermic peak becomes practically constant at 24 h. This means that, at this aging temperature, the material has practically acquired the structural equilibrium at a relative short aging time.

A peak which appears below T_g has been reported for other thermoplastic polymers [8—13] and copolymers [14—15], and thermosetting polymers [3—4, 16]. This peak appears in all cases when the sample is rapidly cooled (quenching in air or in nitrogen liquid) from equilibrium temperature T_0 above T_g, to an aging temperature well below T_g, followed immediately by reheating through the glass transition interval or a short isothermal aging prior to reheating, as has been shown experimentally in polystyrene [17] and in poly(ethylene terephthalate) [18].

This peak is called sub-T_g by some authors [8, 9, 12, 13], and more recently has been called the main peak [17]. The DSC trace of this peak is a consequence of the rapid increase in the enthalpy at a temperature below T_g, followed by a leveling off, as is shown in the path ABCDA in the schematic diagram of enthalpy vs. temperature of Fig. 3. Such behavior is related to the memory effect due to the recovery of enthalpy in the opposite direction to the approach to equilibrium line. This memory effect can be described by phenomenological models taking into account a distribution of the relaxation times [19—23].

Additionally, the endothermic peak superposed on the T_g, which appears in DSC curves of aged samples at 80° and 90°C, can be illustrated in the schematic diagram of enthalpy vs. temperature (Fig. 3). The sample is cooled to T_{a_2} somewhat

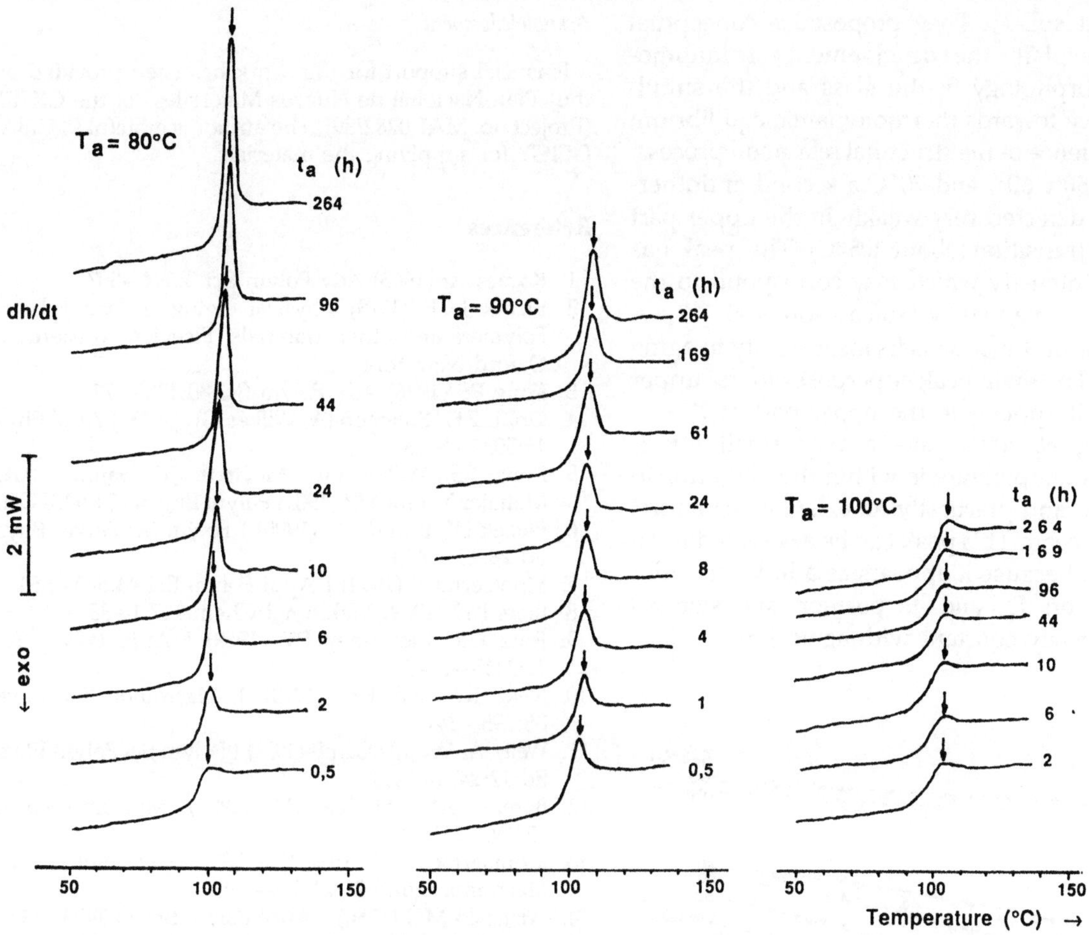

Fig. 2. DSC curves obtained at a heating rate of 10 K/min of an aged epoxy resin at temperatures between 80° and 100 °C for different aging times. The arrows indicate the approximate position of the endothermic peak (or shoulders)

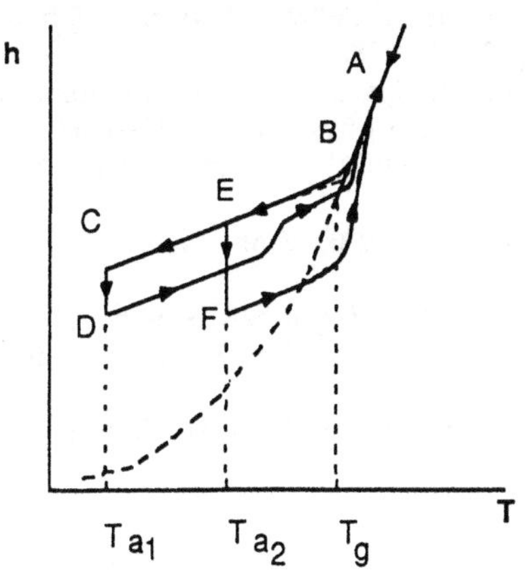

Fig. 3. Schematic diagram of enthalpy vs. temperature illustrating two different thermal histories. a) Cooling, annealing at T_{a_1} well below T_g and reheating (ABCDA path), and b) cooling, annealing at T_{a_2} somewhat lower than T_g and reheating (ABEFA path)

lower than to T_g (ABE path), isothermally annealed (EF path), and reheated above T_g (FA path). In this heating the enthalpy crosses the equilibrium line and immediately moves upwards to recover it again. At this temperature, the endothermic peak appears in DSC curves.

However, the presence of inhomogeneities in crosslinked polymers has contributed to some misinterpretation of the endothermic peak which appears by DSC in aged samples. Kreibich et al. [16] attributed the endothermic peak superimposed on the glass transition to a phase separation of higher and lower crosslinked regions. When increasing the aging teperature, the morphology of the polymer could gradually change to a state where a pronounced short-range orientation takes place. Recently, Mijovic et al. [24] studied the morphological changes in neat epoxies submitted to annealing at sub-T_g. They proposed a conceptual model to explain the development of inhomogeneous morphology in the glass and the simultaneous drive towards thermodynamic equilibrium as a consequence of the structural relaxation process.

For T_a = 50°, 60°, and 70°C a second endothermic peak is detected very weakly in the upper part of the glass transition (about 105°C). This peak has a very low intensity which may correspond to the upper peak, as named by Hutchinson et al. [17]. A characteristic of this peak is its insensitivity to aging conditions. The main peak superposes to the upper peak when it appears in the upper part of T_g.

The aging at 100°C has a very small effect, because this temperature is within the glass transition interval, and practically no extent of structural relaxation is seen. This peak can be associated to an upper peak, because it appears as a little shoulder superposed on T_g, and its position and size remains practically constant with aging time.

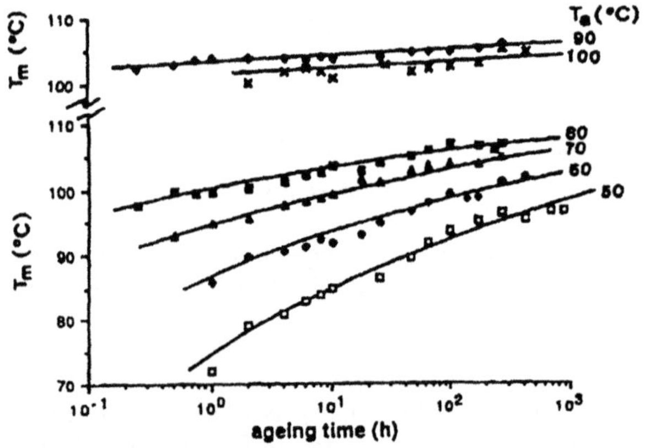

Fig. 4. Representation of the peak temperature vs. aging time for different aging temperatures

The position of the endothermic peak is plotted vs. ($\log t_a$) in Fig. 4. For T_a lower than 90°C these plots are observed not to be linear as pointed out by some authors [6, 8, 9, 12]. Its form shows the characteristic shape of a time-temperature superposition. This figure summarizes the behavior of the above described endothermic peak; the effect of aging time on T_m is greater at low T_a. A higher increase of T_m with $\log t_a$ is observed for T_a between 50° and 70°C, at which sub-T_g peaks are observed, than for T_a = 80°C at which no sub-T_g peaks appear.

Acknowledgement

Financial support for this work has been provided by the "Plan Nacional de Nuevos Materiales" of the CICYT (Project no. MAT 0282/89). The author is grateful to CIBA-GEIGY for supplying the materials.

References

1. Kovacs AJ (1963) Adv Polym Sci 3:394–507
2. Struik LCE (1978) Physical Aging in Amorphous Polymers and other Materials, Elsevier, Amsterdam Oxford New York
3. Kong ES (1986) Adv Polym Sci 80:125–171
4. Ophir ZH, Emerson JA, Wilkes GL (1978) J Appl Phys 49:5032–5038
5. Kong ES, Wilkes GL, McGrath JE, Banthia AK, Mohajer Y, Tant MR (1981) Polym Eng Sci 21:943–950
6. Plazek DJ, Frund ZN (1990) J Polym Sci Polym Phys Ed 28:431–448
7. Montserrat S (1992) J Appl Polym Sci 44:545–554
8. Illers KH (1969) Makromol Chem 127:1–33
9. Foltz CR, McKinney PV (1969) J Appl Polym Sci 13:2235–2245
10. Neki K, Geil PH (1973) J Macromol Sci Phys B8:295–341
11. Weitz A, Wunderlich B (1974) J Polym Sci Polym Phys Ed 12:2473–2491
12. Berens AR, Hodge IM (1982) Macromolecules 15:756–761
13. Montserrat S, Cortés P (1989) Makromol Chem Macromol Symp 20/21:389–395
14. Wizgoski MG (1980) J Appl Polym Sci 25:1455–1467
15. Tant MR, Wilkes GL (1981) J Appl Polym Sci 26:2813–2825
16. Kreibich UT, Schmid R (1975) J Polym Sci Symp No 53:177–185
17. Rudy M, Hutchinson JM (1988) Polym Commun 29:132–134
18. Cortés P, Montserrat S (1989) Makromol Chem Macromol Symp 27:279–287
19. Kovacs AJ, Aklonis JJ, Hutchinson JM, Ramos AR (1979) J Polym Sci Polym Phys Ed 17:1097–1162
20. Narayanaswamy OS (1971) J Am Ceram Soc 54:491–498
21. Moynihan CT et al. (1976) Ann NY Acad Sci 279:15–35
22. Rekhson SM (1986) J Non Cryst Solids 84:68–85
23. Hodge IM, Berens AR (1982) Macromolecules 15:762–770
24. Mijovic J, Lin K-F (1986) J App Polym Sci 32:3211–3227

Author's address:

Dr. S. Montserrat Ribas
Universitat Poltècnica de Catalunya
E.T.S. Enginyers Industriales de Terrassa
Carrer Colon 11
08222-Terrassa, Spain

Oriented polymers

Crystallization in stretched and unstretched rubber-thermodynamics and kinetics

R. N. J. Conradt, B. Heise, and H. G. Kilian

Abteilung Experimentelle Physik, Universität Ulm, F.R.G.

Abstract: Melting as well as the time dependence of thermally and strain-induced crystallization and crystallization-induced relaxation in networks are described within the framework of a single theory. The thermodynamics of eutectoid copolymers is combined with the van der Waals network model and with a new formulation of nucleation in eutectoid multi-component systems. The total energy balance is considered. The physics behind the mean-field approach as well as its shortcomings are discussed.

Key words: Net work; thermally or strain-induced crystallization; melting; thermodynamics; kinetics

Introduction

Stereoregularity of polymer chains is a necessary premise for crystallization. It is clear that junctions in networks are non-crystallizable. Networks therefore behave like random copolymers, the co-units of which cannot be built in the same lattice. It is then reasonable to consider network-chains of different lengths as different components of a multi-component system. It is now a crucial point that mixed extended-chain crystals are formed with a maximum disparity of chain lengths. A network with a broad chain-length distribution therefore crystallizes like a eutectoid multi-component system. Primary structure elements like crystallinity or crystal thickness distribution are determined by the chemical structure of chains. Chain segregation of different chain lengths is a strong thermodynamic condition that influences the kinetics of crystallization.

In order to describe strain-induced crystallization, one has to formulate the strain energy in a stretched network. This is achieved with the aid of the van der Waals network approach. By combining this theory with the thermodynamics of eutectoid copolymers [1], strain-induced crystallization can be described. Here, it is the most important point that the intrinsic strain in the non-crystallized region is correctly defined. In this manner the stress-strain hysteresis of natural rubber [2] was interpreted as a sequence of constrained equilibrium-states [1]. Yet, the parameters introduced are difficult to connect with a precise interpretation.

It was clear that progress in understanding can only be achieved by developing an adequate treatment of crystallization kinetics in networks. Such an approach is offered in this paper. It is shown that heterogeneous nucleation without crystal growth is sufficient to interpret the experiments. Known concepts [3, 4] are extended so that nucleation in eutectoid multicomponent systems can be described. Changes of the partial molar Gibbs-free-enthalpy by mixing or demixing processes, as well as by deformation, play a decisive role. The understanding is substantially based on the correct formulation of thermodynamic factors.

The theoretical approach is tested by studying crystallization of radiation crosslinked polyethylene and of natural rubber. Stress-strain cycles are carried out whereby the whole energy balance is measured in a Müller-stretching micro-calorimeter [1, 5—10]. These and other experiments like crystallization-induced stress-relaxation are discussed in quantitative terms.

Theory

We have first to describe the equilibrium states in unstretched or stretched semicrystalline networks.

The network as eutectoid copolymer

Crosslinks are lattice-incompatible. The network is therefore first characterized by the mole fraction of the crystallizable units (*c*-units) [1, 11–14],

$$x_c = \frac{n_c}{n_c + n_{nc}} ; \tag{1}$$

n_c and n_{nc} are the mole numbers of the *c*-units and the non-crystallizable units (*nc*-units; each junction includes two *nc*-units!). Together with the mole fraction of the *nc*-units, we have

$$x_c + x_{nc} = 1 . \tag{2}$$

Chains of different lengths are considered as components of the system. The networks studied in this paper have a random chain-length distribution [1, 11, 12]. The mole fraction of chains with length y is then explicitly given by

$$x_y = x_{nc} x_c^{y-1} \tag{3}$$

with the normalization condition

$$\sum_{y=1}^{\infty} x_y = 1 . \tag{4}$$

By squeezing the junctions out of the core, extended-chain mixed crystals (EMCs) are formed. A two-dimensional sketch of an EMC is shown in Fig. 1. Defects in the front layers limit the solubility. Each of the EMCs is defect-saturated characterized by a maximum disparity of chain-lengths Δy. Internal equilibrium guarantees that the front layers have a well defined thickness which must be related to Δy. The maximum disparity of chain lengths in EMCs is written [15]

$$\Delta y = Ay + B \; ; \quad B = \frac{x_c}{x_{nc}} B_1 + B_2 \, , \tag{5}$$

where y is the mean chain-length in the EMC. A, B_1, and B_2 are constants.

If Δy is much smaller than the width of the chain-length distribution the thickness distribution of EMCs is determined by the chain-length distribution itself. Hence, the kinetics of crystallization include segregation of components. The kinetics are strongly controlled by thermodynamic factors.

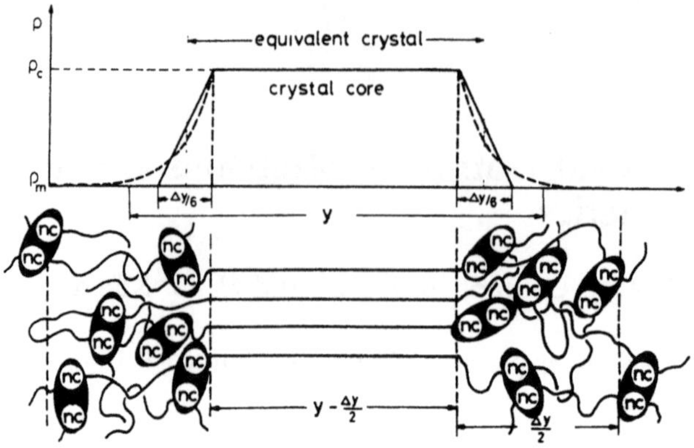

Fig. 1. Sketch of a cross-section of an inhomogeneous microphase. The junctions are schematically indicated by pairs of *nc*-units. At the top is a representation of the mean density (ρ_c crystal-density, ρ_m density of the rubbery matrix). y is the mean chain length in th EMC. Δy is the maximum disparity of chain-length in the EMC

Primary parameters of the colloid-structure

Any attempt at describing strain-induced crystallization requires at least some knowledge about primary structure parameters like the degreee of crystallinity, the average thickness, and the thickness distribution of EMCs, as well as the mean of the chain length in the non-crystallized regions. For eutectoid copolymers they are completely determined by the chain-length distribution [1, 11–14]. For "random networks" the primary parameters can be explicitly formulated.

Because the smallest EMCs with the mean thickness $y_u(T)$ are always the ones which are just stable at a given temperature, the primary colloid parameters must be functions of $y_u(T)$. The molar fraction of chains built in EMCs related to their total number, w_p, is written

$$w_p(T) = \frac{\sum_{y=y_u(T)}^{\infty} y x_y}{\sum_{y=1}^{\infty} y x_y} = x_c^{y_u(T)} [y_u(T) x_{nc} + x_c] \, . \tag{6}$$

Accounting for those *c*-units which are built by the defect-layers, we define parameter δ:

$$y_{\text{cryst}} = y - \delta \Delta y \, . \tag{7}$$

The molar degree of crystallinity $w_c(T)$ is then written as

$$w_p(T) = \frac{\sum_{y=y_u(T)}^{\infty} (y - \delta \Delta y) x_y}{\sum_{y=1}^{\infty} y x_y}$$

$$= x_c^{y_u(T)} [(y_u(T) - \delta \Delta y_u(T)) x_{nc} + x_c] \; . \quad (8)$$

$\Delta y_u(T)$ is the same as in Eq. (7), provided y_u replaces y. If pair-interaction is dominant δ should be equal to 1/3 [16]. With the parameter y_k:

$$y_k = \frac{B}{3 - A} , \quad (9)$$

the special formulation of the molar degree of crystallinity results in

$$w_c = \left(1 - \frac{A}{3}\right) x_c^{y_u(T)} [(y_u(T) - y_k) x_{nc} + x_c]$$

$$\langle y^{(c)} \rangle = \frac{\sum_{y'=y_u(T)}^{\infty} \left(1 - \frac{A}{3}\right) (y' - y_k) x_{y'}}{\sum_{y'=y(T)}^{\infty} x_{y'}} \quad (10)$$

$$= \left(1 - \frac{A}{3}\right) \left(\frac{x_c}{x_{nc}} + y_u(T) - y_k\right) .$$

The molar crystallinity w_c does not, in any case, describe the influence of "crystals" on a mean property of a semicrystalline system. At large deformations of semicrystalline polyethylene, one finds, for example, a better understanding if one assumes that the EMCs including the defect-layers act as solid filler. In this case, one has to use w_p instead of w_c.

The mean length of crystallized c-sequences is written as

$$\langle y^{(p)} \rangle = \frac{\sum_{y'=y_u(T)}^{\infty} y' x_{y'}}{\sum_{y'=y_u(T)}^{\infty} x_{y'}} = \frac{x_c}{x_{nc}} + y_u(T) \; . \quad (11)$$

The "two-phase" thickness of EMCs reads accordingly

$$\langle y^{(c)} \rangle = \left(1 - \frac{A}{3}\right) \frac{\sum_{y'=y_u(T)}^{\infty} (y' - y_k) x_{y'}}{\sum_{y'=y_u(T)}^{\infty} x_{y'}}$$

$$= \left(1 - \frac{A}{3}\right) \left(\frac{x_c}{x_{nc}} + y_u(T) - y_k\right) . \quad (12)$$

The coexistence temperature

EMCs of the same thickness are thermodynamically equivalent. This holds true independent of where they are located (autonomy of the EMCs). If the solubility of each of the EMCs is defined as in Eq. (5), it is possible to deduce an explicit relation for the temperature of coexistence T_{yu}, i.e., that temperature at which the chemical potentials with chain length y_u in the melt and in the EMC are equal. The coexistence temperature T_{yu} of smallest EMCs with the thickness $y_u(T)$ is then written as [1, 11, 14]

$$T_{yu} = T_M \frac{1 - 2\sigma_e / N_1}{N_2} , \quad (13)$$

with

$$N_1 = \left(1 - \frac{A}{3}\right) \Delta h(T_{yu}) (y_u(T_{yu}) - y_k) \quad (14)$$

$$N_2 = 1 + \frac{R T_M}{N_1} \left[\ln \left(\frac{y_y(T_{yu})}{\Delta y_u(T_{yu})/2 + 1} \right) - \ln \left(\frac{x_{yu}^{(m)}}{x_{yu}^{(c)}} \right) + \frac{y_u(T_{yu}) - 1}{2} \chi^{(m)} \right] , \quad (15)$$

where T_M is the melting temperature of the extended chain crystal with infinite thickness. $\Delta h(T)$ is the molar melting enthalpy per c-unit.

$$\Delta h(T) = \Delta h(T) - \Delta C (T_M - T) \; . \quad (16)$$

ΔC is the difference between the molar heat capacities in the melt and in the crystal. The ratio of the surface free energy $2\sigma_e(T)$ over $\Delta h(T)$ is independent of temperature. $\chi^{(m)}$ defines the mean interaction parameter per c-unit in the multi-component melt. This excess entropy arises because the chains are linked in a network.

The mole fraction of the c-sequences with the length y_u in the melt related to the mole fraction in EMCs of the thickness y_u is equal to

$$\frac{x_{yu}^{(m)}}{x_{yu}^{(c)}} = \frac{(1-\varepsilon)[x_c^{-\Delta y_u(T)/2} - x_c^{\Delta y_u(T)/2}]}{(1-x_c^{y_u(T)})[1-x_c^{\Delta y_u(T)/2}} \qquad (17)$$

with

$$\varepsilon = \frac{1-x_c^{\Delta y_u(T)/2}}{x_c^{-\Delta y_u(T)/2} - x_c^{\Delta y_u(T)/2}} \,. \qquad (18)$$

Equation (13) is an implicit equation in T_{yu}, so that one has to solve it numerically through an iteration procedure.

It is a very important feature that ECMs of different thickness melt consecutively, starting at the lowest temperature with the smallest EMCs, and ending with the fraction of largest EMCs at the maximum melting temperature [11, 12]. Moreover, most of EMCs have to adjust their mass fraction and their intrinsic concentration. A continuous reorganization of the micro-structure is therefore possible.

It is a crucial point whether crystallization behaves in the same manner, starting with largest EMCs and ending with the smallest ones, always controlled by the mass-balance. This can only be verified if the crystallization kinetics in networks are described appropriately.

Now, let us write the difference between the chemical potentials of c-sequences of the length y_u in the melt and in the EMC, $\Delta\mu_{yu\text{-mix}} = \mu_{yu\text{-mix}}^{(m)} - \mu_{yu\text{-mix}}^{(c)}$ at temperatures $T < T_{yu}$ in the metastable range. The composition in the phases is taken to be practically the same as at the coexistence temperature T_{yu}. It is assumed, in addition, that the "equilibrium maximum disparity of chain lengths" is not changed in the metastable regime (see Eq. (5)). If we introduce the approximation

$$\Delta g(T) = \Delta h(T) - T\Delta S(T) \approx \Delta h(T)\left(1-\frac{T}{T_M}\right);$$

$$\Delta S(T) \approx \frac{\Delta h(T)}{T_M}\,, \qquad (19)$$

we finally arrive at

$$\Delta\mu_{yu\text{-mix}} = \left(1-\frac{\Delta y_u(T)}{3}\right) \cdot \Delta h(T)\left(1-\frac{T}{T_M}\right) - 2\sigma_e$$

$$- RT\left[\ln\left(\frac{y_u(T_{yu})}{\Delta y_u(T_{yu})/2+1}\right) - \ln\left(\frac{x_{yu}^{(m)}}{x_{yu}^{(c)}}\right) + \frac{y_u(T_{yu})-1}{2}\chi^{(m)}\right]. \qquad (20)$$

The above conditions need not strictly hold true when nuclei as mixed micro-phases are formed at temperatures below T_{yu}. Equation (20) will nevertheless be used for describing nucleation in strained and unstrained networks.

The strain-energy of a van der Waals network

If we want to understand strain-induced crystallization, we have to know the strain-energy in a stretched network. In the model of a van der Waals network the strain energy function is explicitly known [19, 20].

To deduce this, let us first define the deformation function Φ:

$$\Phi = \frac{1}{2}\left(\sum_{k=1}^{3}\lambda_k^2 - 3\right) = \frac{1}{2}(I_1 - 3)\,, \qquad (21)$$

where I_1 is the first strain-invariant. The strain-energy of the Gaussian network [2, 17, 18] is then equal to

$$W_{\text{gauss}} = G\Phi\,, \qquad (22)$$

whereby the modulus G is written as

$$G = \frac{\rho RT}{M_u\langle y^{(m)}\rangle}\frac{\langle r^2\rangle}{\langle r_0\rangle}\,, \qquad (23)$$

where ρ is the density, R the gas constant and T the absolute temperature. $\langle r^2\rangle/\langle r_0^2\rangle$ is the memory term. The mean chain length in a semicrystalline network is defined by

$$\langle y^{(m)}\rangle = \frac{\sum_{y'=1}^{y_u(T)-1} x_{y'} y'_\zeta}{\sum_{y'=1}^{y_u(T)-1} x_{y'}}\,; \quad y'_\zeta = \frac{M_{\text{unit}}}{M_u}\,y\,; \qquad (24)$$

y_ζ gives the number of strain-invariant units per chain expressed as number of periodic units. M_u is the molecular weight of the stretching invariant unit that should not be identical with the monomer unit. The stretching invariant segment is an entropy-invariant unit that is always smaller than the Kuhn-segment [1, 2, 12, 17]. In non-crystallized networks $\langle y^{(m)}\rangle$ is, of course, identical with mean chain length $\langle y_0^{(m)}\rangle (y_u(T) = \infty)$.

The van der Waals model is now a modification of the Gaussian network by accounting for finite chain extensibility and global interactions [19, 20]. The first parameter is the maximum strain $\lambda_{max,0}$ which is related to the mean chain-extensibility in the network according to

$$\lambda^2_{max,0} + \frac{2}{\lambda_{max,0}} = \sum x_y \sqrt{y_\zeta} = \langle\sqrt{y_\zeta}\rangle$$

$$\lambda_{max,0} \gg \frac{2}{\lambda_{max,0}}; \; \lambda_{max,0} = \sqrt{\langle\sqrt{y_\zeta}\rangle}\,. \tag{25}$$

This formulation implies that chains of different lengths store the same amount of strain energy. Chains of different lengths are individually deformed so as to guarantee equipartition of strain-energy. At the same macroscopic strain λ_{max} chains of different lengths are fully extended to $\lambda_{max,y} = (y_\zeta)^{1/2})$.

The second van der Waals parameter a was shown to be related to the junctions fluctuation [21–26].

The strain-energy of a van der Waals-network is then formulated as

$$W = -2G\left(\phi_{max}[\ln(1 - \sqrt{\phi/\phi_{max}}) + \sqrt{\phi/\phi_{max}}] + \frac{1}{3}a\phi^{3/2}\right). \tag{26}$$

This energy is characterized by a pole at $\Phi = \Phi_{max}$. The term which is proportional to the second van der Waals parameter a enforces mechanical softening due to global interactions between the chains.

In the mode of simple extension the deformation function Φ is equal to

$$\phi = \frac{1}{2}\left(\lambda^2 + \frac{2}{\lambda} - 3\right), \tag{27}$$

with λ as the macroscopic strain in direction of the force. Φ_{max} is then written

$$\phi_{max} = \frac{1}{2}\left(\lambda^2_{max,0} + \frac{2}{\lambda_{max,0}} - 3\right). \tag{28}$$

From Eq. (22), we can obtain the nominal force f:

$$f = \left(\frac{\partial W}{\partial \lambda}\right)_{T,p} = GD\left(\frac{1}{1-(\phi/\phi_{max})^{1/2}} - a\phi^{1/2}\right), \tag{29}$$

with

$$D = \lambda - \lambda^{-2}. \tag{30}$$

According to Eq. (25) the non-Gaussian features become more important if the chains are shorter. This is significant in semicrystalline networks because $\langle y^{(m)}\rangle$ decreases with crystallization becoming, finally, relatively short (see Eq. (24)).

It is important to prove whether network chains are still energy-equivalent in the presence of crystals. This would be verified if EMCs do not behave like reinforcing filler particles [27].

The matrix strain

We now give a mean-field formulation of the intrinsic strain in the amorphous matrix in a strained and semicrystalline network. An important criterium here is that force-extension curves of a semicrystalline system reflects the deformation in the rubbery regions only [25]. We can therefore be sure of understanding the stress-strain pattern of crystallizing networks if the intrinsic strain in the rubbery regions is correctly defined. We confine ourselves to a mean field Ansatz in the hope that this approach works during crystallization as well.

By finite element calculations a complicated stress-strain pattern is obtained. But, a comparison with the mean field approach defends its reliability.

It has been shown by many authors that crystals formed in a network under strain are fully oriented [1–3, 28–38]. Therefore, the intrinsic strain in the matrix λ_i must fall below the macroscopic strain λ [1, 2, 18]. This crystallization-induced shrinkage was first discussed by Flory [39, 40].

In an unstretched network the distance between junctions should approxiamtely be equal to $\langle y_0^{(m)}\rangle^{1/2}$. In a non-crystallized stretched system this strain increases proportionally to $\lambda(\langle y_0^{(m)}\rangle)^{1/2}$. When fully oriented EMCs are formed, the mean intrinsic strain λ_i decreases. The decrease of the intrinsic strain is determined by the mean effective thickness of the total number of EMCs $x^{(c)}$. From the sketch in Fig. (2), we are led to formulate the relationship

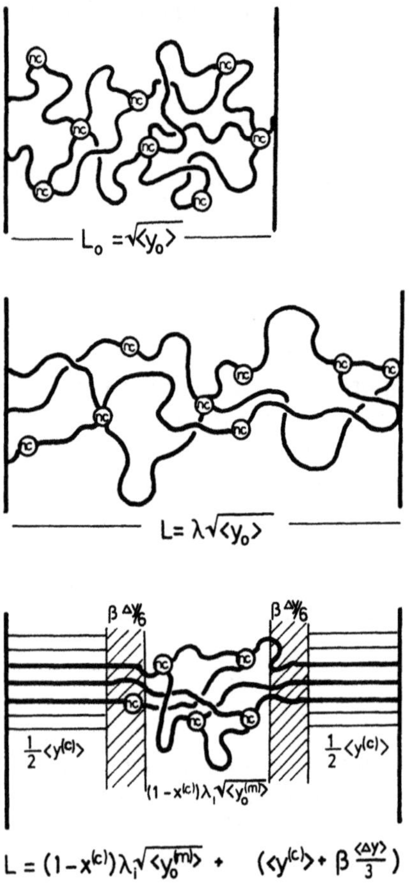

Fig. 2. Shrinkage of the non-crystallized chains due to oriented crystallization of EMCs

$$\lambda_i = \frac{\lambda\sqrt{\langle y_0^{(m)}\rangle} - \left(\langle y^{(c)}\rangle + \beta\,\frac{\langle \Delta y\rangle}{3}\right)x^{(c)}}{\sqrt{\langle y^{(m)}\rangle}\,(1 - x^{(c)})}$$

$$\langle y_0^{(m)}\rangle = \frac{x_c}{x_{nc}}\,; \quad \langle \Delta y\rangle = A\langle y^{(p)}\rangle + B \tag{31}$$

$$\langle y^{(m)}\rangle = \frac{\sum_{y'=1}^{y_u(T,\lambda)-1} y' x_{y'}}{\sum_{y'=1}^{y_u(T,\lambda)-1} x_{y'}} = \frac{(1 - x_c^{y_u(T,\lambda)})/x_{nc} - (y_u(T,\lambda) - 1)\,x_c^{y_u(T,\lambda)-1}}{1 - x_c^{y_u(T,\lambda)-1}}\,.$$

β defines the "rigid part" in the boundaries which is fully linked to the crystal core. $x^{(c)}$ as the total number of crystallized chains is given by

$$x^{(c)} = x_{nc}\,x_c^{y_u(T,\lambda)}\,. \tag{32}$$

The smallest stable EMC with the thickness $y_u(T,\lambda)$ is now a function of temperature and strain. λ_i therefore depends on $w_p(T,\lambda)$, $\langle y^{(p)}(T,\lambda)\rangle$ and $\langle y^{(m)}(T,\lambda)\rangle$ in a complicated manner.

The elastic component of the chemical potential

The elastic term in the chemical potential of chains is easily deduced if strain-energy is equipartioned. The strain-energy term in the chemical potential is then identical for chains with different lengths. The coexistence temperature of EMCs (the melting of which can be observed) is lifted with strain. The relative shift is larger, the smaller the thickness of EMCs.

When the chains in the regime $y \geqslant y_u(T)$ are crystallized the strain energy in the multi-component melt should be rewritten (Eq. (22)):

$$W = RTw_\Sigma = RTw(\lambda_i, \lambda_{\max}(\lambda), a)\sum_{y=1}^{y_u(T,\lambda)-1} n_y^{(m)}\,. \tag{33}$$

The strain energy is proportional to $\Sigma n_y^{(m)}$, which is the total mole number of chains in the rubbery regions [20]. Because of the selective segregation of chains in EMCs the maximum strain $\lambda_{\max}$ of the rubbery regions decreases with increasing crystallinity. $\lambda_{\max}(\lambda)$ is defined by

$$\lambda_{\max}(\lambda) = \frac{\sum_{y'=1}^{y_u(T,\lambda)-1} x_{y'}\sqrt{y'_\zeta}}{\sum_{y'=1}^{y_u(T,\lambda)-1} x_{y'}}\,. \tag{34}$$

We are now in the position to deduce the elastic component of the chemical potential of chains with the length $y_u(T,\lambda)$ in the melt of a semicrystalline and stretched network

$$\lambda_{yu\text{-elast}}^{(m)} = RT\left[\frac{\partial w_\Sigma}{\partial n_y^{(m)}}\right]_{T,p} = RT\left[w + \frac{\partial w_\Sigma}{\partial \lambda_i}\,\frac{\partial \lambda_i}{\partial n_y^{(m)}} + \frac{\partial w_\Sigma}{\partial \lambda_{\max}}\,\frac{\partial \lambda_{\max}}{\partial n_y^{(m)}}\right]_{T,p}. \tag{35}$$

The derivatives are given in the Appendix.

The elasto-chemical potential

It is assumed that the "chemical-" and strain energy components can be added together. With the aid of Eqs. (20) and (35) the difference of the elasto-chemical potential in the melt and the chemical potential in the EMCs is then simply written as

$$\Delta\mu_{yu} = \Delta\mu_{yu\text{-mix}} + \mu^{(m)}_{yu\text{-elast}} \,. \tag{36}$$

The differential heat-exchange

One of the best identifications of strain-induced crystallization is the measure to total energy-balance during the stretch in a stretching micro-calorimeter [1]. This is true because crystallization- or melting-heats overpower thermo-elastic effects in the rubbery regions [1, 2]. It is a severe test of the theory developed to determine whether the differential heats exchanged during deformation are well described.

Let us first consider the thermodynamic limit. To this end, we formulate the differential of the Gibbs-function $G(T, p, L)$

$$dG = -SdT + Vdp + FdL \,, \tag{37}$$

with $S(T, ,p, L)$ as the entropy of the system, $F(T, p, L)$ is the applied force. According to the conditions of integrability, thermal and elastic properties in the rubbery matrix are uniquely interrelated. The differential heat $(\partial Q/\partial\lambda)_{T,p}$ exchanged during a quasi-isothermal stretch at the temperature T, including the heats of melting or crystallization, can be written as

$$\left[\frac{\partial Q}{\partial\lambda}\right]_{T,p} = T\left[\frac{\partial f}{\partial T}\right]_{L,p} + \Delta h(T)\left[\frac{\partial w_c}{\partial\lambda}\right]_{T,p} \,; \tag{38}$$

f is defined as the nominal force with Eq. (29) whereby the modulus, as in Eq. (23), may be reformulated by expressing the temperature dependence of the memory term in a linear approximation:

$$G_r = G e^{(2\beta-\mu)(T-T_0)} \,. \tag{39}$$

T_0 is the temperature of reference at which the network has been formed. We rest content in this work by setting this temperature equal to T_0 = 273 K. β is the linear coefficient of thermal expansion:

$$B = \frac{\partial \ln(L_0(T))}{\partial T} \,. \tag{40}$$

For non-iso-energetic rotational isomers the temperature coefficient of the mean square length of the equivalent free chain μ changes according to

$$\mu = \frac{\partial \ln(\langle r_0^2(T)\rangle)}{\partial T} \,. \tag{41}$$

The mean square average of the junction-to-junction distance $\langle r^2\rangle$ was taken to depend on the change in volume according to

$$2\beta = \frac{\partial \ln(\langle r^2(T)\rangle)}{\partial T} \,. \tag{42}$$

The above equations (including the ones given in the Appendix) describe the thermo-elastic phenomena in the rubbery regions as well as with crystallization or melting. Thermo-elasticity in the crystals is neglected due to the low thermal expansion, which is one order of magnitude lower then the one of the rubbery regions, and because of having a maximum degree of crystallinity that is lower than $w_c \approx 0.2$.

The nucleation-rate in eutectoid polymers

In eutectoid multi-component systems nucleation-induced "fractionation" of chains of different lengths has to occur so as to finally arrive at the EMC-thickness distribution as demanded by thermodynamics. This should hold true in both modes of crystallization, thermally and strain-induced crystallization.

In contrast to Treloar [2], we postulate that EMCs do not behave like active fillers. The thermodynamic reason is that most of the EMCs have to adjust, at each temperature, their composition as well as their mass fraction. Due to necessary reshufflings, it is not very likely that EMCs behave like rigid fillers.

We make the following assumptions:

a) Only nucleation is considered.
b) EMCs should be defect-saturated. The maximum disparity of chain lengths is fixed by very large defect energies. This might be taken as a reason for assuming that the composition in EMCs cannot be modified, even in metastable states.

c) The multi-component melt is treated as a homogeneous matrix.

In these circumstances, strain-induced nucleation of EMCs of the thickness y_u should depend on the chemical potentials of the chains with the length y_u in the metastable melt, and in EMCs at temperatures T below $T_{yu}(\lambda)$, the temperature coexistence. Oriented EMCs are stable at temperatures above the maximum melting point of a non-stretched network because the entropy in the melt is reduced.

It is reasonable to assume that the shape of the nuclei is cylindrical [4, 40]. A sketch of such an EMC-nucleus is shown in Fig. 3. Because longitudinal defect-layers are built up, even in the first steps, longitudinal growth is not possible at all. The stable nucleus in a stretched state is thus characterized by the critical radius r^*.

The difference of the chemical potentials on an EMC-embryo with the thickness $y_u(T, \lambda)$ and the radius r is equal to

$$\Delta\mu_{yu\text{-nuc}} = r^2 \pi \Delta\mu_{yu} - 2r\pi \left(y_u(T, \lambda) - \frac{\Delta y_u}{3} \right) \sigma_M . \quad (43)$$

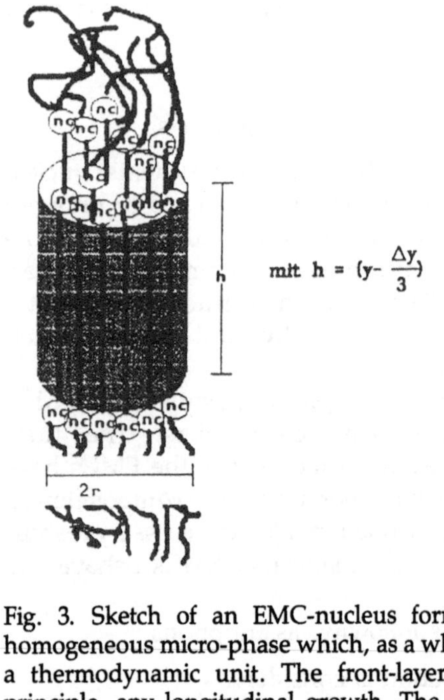

Fig. 3. Sketch of an EMC-nucleus formed as an inhomogeneous micro-phase which, as a whole, represents a thermodynamic unit. The front-layers suppress, in principle, any longitudinal growth. The stability of the nucleus is substantially dependent on the relative thickness of the longitudinal defect-layers. As an approximation, it is assumed that the mean thickness and the composition in the mixed micro-phase there with the defect-energies in the front layers of the nucleus are the same as in equilibrium conditions

Hence, the stability of EMC-nuclei is substantially influenced by the defects in the front-layers. σ_M is the free surface enthalpy of the lateral surface areas which should be in the magnitude of order as reported in literature [4]. One has to be aware that the "thermodynamic driving force" to from an EMC-nucleus is controlled by the difference between the temperature of coexistence T_{yu} and the actual temperature T. Because the composition and the strain in the melt change with crystallization, T_{yu} is shifted when crystallization proceeds (see Eq. (13)). The degree of undercooling or overdrawing is therefore continuously modified during crystallization. This is a consequence of a complicated segregation of chains in a network with a broad chain-length distribution.

The critical radius r^* of the nucleus is deduced by minimizing $\Delta\mu_{y\text{-nuc}}$ (Eq. (41))

$$r^* = \frac{[y_u(T, \lambda) - \Delta y_u/3]\sigma_M}{\Delta\mu_{yu}} . \quad (44)$$

The critical free enthalpy of a stable nucleus is thus equal to

$$\Delta\mu^*_{yu} = \frac{(y_u(T, \lambda) - \Delta y_u/3)^2 \sigma_M^2 \pi}{\Delta\mu_{yu}} . \quad (45)$$

It is not easy to define how heterogeneous nucleation depends on the concentration of the chains $x_y^{(m)}$. Modifying conventional relations [4], we define the nucleation rate I_y by

$$I_y = x_y^{(m)}(t) \frac{RT}{\Delta h(T)} \{f_1 + (y(x_{yo}^{(m)} - x_y^{(m)}(t)))^n f_2\} \cdot e^{-\frac{\Delta\mu^*_{yu}}{RT}} e^{-\frac{\Delta G_\eta}{R(T-T_G)}} , \quad (46)$$

where the total number of chains with the length y, $x_{yo}^{(m)}$, is equal to (see Eq. (3))

$$x_{yo}^{(m)} = \frac{x_y}{1 - x_c^{(y_u((T,\lambda)-1)}} . \quad (47)$$

Under the simplest circumstances, formation of a stable nucleus should be proportional to the momentary relative number of c-sequences in the melt $x_y^{(m)}(t)$ (Eq. (3)). $x_y^{(m)}(t)$ gives the fraction of chains that are not yet built in EMCs.

The exponential term in Eq. (46), including ΔG_η, accounts for changes of the melt-viscosity. T_G is the quasi-static glass transition temperature, f_i are empirical constants. The term which is proportional to f_2, describes an influence of existing EMCs. The larger the thickness of these EMCs, the more nucleation is supposed to become enhanced. The power-law coefficient n characterizes the cooperative nature of these effects.

The rate of nucleation should depend on temperature in the traditional manner. One expects a maximum at a well defined temperature above T_G. On the other hand, it is interesting to study how the quasi-isothermal nucleation rate is changed by overdrawing. A complicated concert of simultaneously forming EMCs with different thicknesses is expected for strain-induced crystallization. Each T_y is changed during crystallization, modifying undercooling or overdrawing accordingly. It is, in principle, not guaranteed that nucleation always starts with largest EMCs and ends up with the smallest ones. We will come back to this point later.

Experimental

We have investigated natural rubber and polyethylene crosslinked to different degrees (Tables 1 and 2).

Table 1A. Natural rubber

$x_{nc} = 0.02$
$\rho/(M_u \lambda^2_{max}) = 1.06\ 10^{-3}$ MPaK^{-1}; $\lambda_{max,0} = 10.5$; $a = 0.287$
$f_1 = 0.0115$; $f_2 = 11.5$; $n = 0.66$:
thermally induced crystallization
$f_1 = 1.15$; $f_2 = 11.5$; $n = 0.6$:
strain-induced crystallization
$\Delta G_\eta = 1550$ J mol^{-1}; $T_G = 209$ K

Table 1B. Natural rubber — general data

$T_m = 298$ K; $\Delta h(T_m) = 4350$ J mol^{-1};
$\Delta C = 6$ J mol^{-1} K^{-1};
$2\sigma_e/\Delta h = 2.06$; $\sigma_m = \Delta h(T_m)/20$
$A = 0.2$; $B_1 = 3.5$; $B_2 = 40$; $a = 1.3$; $\chi^{(m)} = 0.1$
$G(T) = (\rho RT)/M_u$; $M_u = 68$ g mol^{-1};
$\rho(T_0) = 0.96\ 10^3$ kg m^{-3}; $T_0 = 273$ K
$\beta = 2.2\ 10^{-4}$ K^{-1}; $\mu = 9.6\ 10^{-4}$ K^{-1}

Table 2A. Thermodynamic parameters of LDPE

$T_M = 415$ K; $\Delta h(T_M) = 4046$ Jmol^{-1},
$\Delta C = 6$ Jmol^{-1}; $2\sigma_e/\Delta h = 2.1$
$A = 0.15$; $B_1 = 0.9$, $B_2 = 80$

Table 2B. Radiation-crosslinked polyethylene

$a = 1.3$; $n = 0.66$

100 Mrad	$x_{nc} = 0.026$; $\chi^{(m)} = 0.05$ $f_1 = 5\ 10^{-4}$, $f_2 = 1.5$
400 Mrad	$x_{nc} = 0.067$, $\chi^{(m)} = 0.2$ $f_1 = 1\ 10^{-4}$, $f_2 = 0.3$

WAXS-investigations are carried out mainly with synchrotron-x-rays in DESY.

A Perkin-Elmer device was used for measuring temperature-induced melting and crystallization under constant heating rate.

Stress-strain pattern has been measured in a Zwick-stretching apparatus. The degree of extension was determined by optical methods. Caloric

FLOW-DIAGRAM OF THE COMPUTER PROGRAM

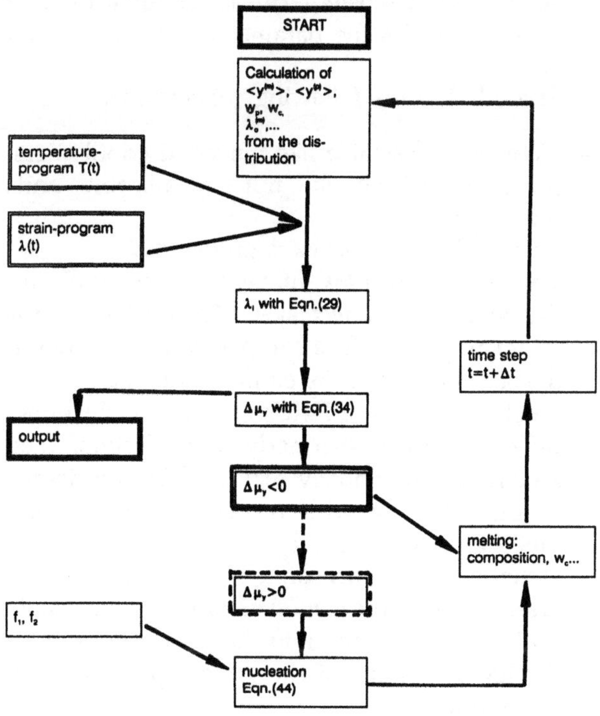

effects during the stretch under varying conditions (constant strain-rate or discrete increments of strain) were observed in a stretching micro-calorimeter of the Müller type [1, 5—10].

To calculate the melting curves or the strain-induced crystallization a computer program was developed (see flow diagram). In metastable states, under a defined undercooling or overdrawing, those fractions are picked out which are able to form stable nuclei. First, the nucleation rate is computed up to a relatively small number of EMCs, then thermodynamic factors are adjusted and the next calculus is run through. This kind of calculation is continued until the system is near to equilibrium. It is possible to follow temperature-induced or quasi-isothermal strain-induced crystallization with the same program.

Results and discussion

Temperature-induced crystallization of polyethylene-networks

Heating or cooling of two differently crosslinked polyethylenes with constant rate yields DSC-diagrams as depicted in Fig. 4. The solid lines are calculated by using the parameters collected in Table 2B. The "viscosity term" is set equal to 1. Its influence modifies the parameters f_i by a constant factor.

We are led to the following conclusions:

- Networks crystallize like eutectoid copolymers. Junctions behave like non-crystallizable units (*nc*-units).
- Segregation of *nc*-units is guaranteed.
- The stability of EMC-nuclei is substantially influenced by the defect layers. The nuclei are mixed micro-phases. The composition seems not to deviate much from its equilibrium value.
- Nucleation rates of EMCs of different thickness differ by their thermodynamic factors. The kinetics is individually regulated by the degree of undercooling with respect to the varying melting temperatures T_{yu}.
- Heterogeneous nucleation is dominant.
- Heterogenous nucleation is enhanced if stable EMCs exist. According to the f_2-term in Eq. (44) the EMCs volume of the same type should be important.

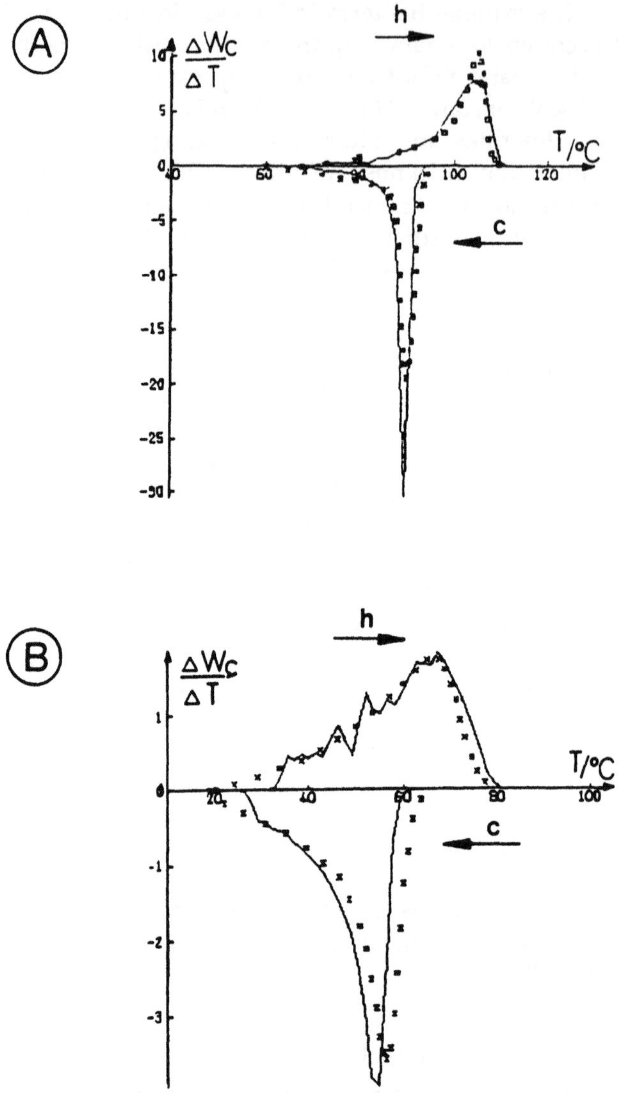

Fig. 4. DSC-traces of radiation-crosslinked PE-networks at a heating or cooling rate of 5 K min^{-1} A) 100 Mrd and B) 400 Mrd. The solid lines are computed by using the parameters in Table 2

- Nucleation seems to be controlled by the momentary number of the *c*-sequences available in the melt ($I_y \approx x_y^{(m)}(t)$).
- Nucleation should occur on very local scales.
- During cooling, nucleation of EMCs starts with largest EMCs, then followed by smaller ones. Simultaneous nucleation of EMCs with different thickness is consecutive in so far as EMCs grow slower the smaller they are.
- Crystal growth is not important as long as folded-chain crystallization does not occur.

The f_is of the differently crosslinked polyethylenes both differ by the same factor of about 5. The reason remains unknown; perhaps it results from a different "viscosity".

Temperature-induced crystallization and melting of natural rubber

We would like to verify our approach by investigating crystallization of natural rubber. Natural rubber was isothermally crystallized at T = 248 K for 24 h. The DSC-traces of a stretched semicrystalline sample that is kept at a constant length are not fully reproduced by "equilibrium melting curves". This is shown in Fig. 5A. But, when we exactly examine the experimental conditions by applying our kinetic approach, there is no longer any disagreement (Fig. 5B). Apparently, some of the EMCs in natural rubber need a very long time to nucleate [2].

Isothermal crystallization of natural rubber

As shown in Fig. 6 the quasi-isothermal crystallization of natural rubber at defined undercoolings is fairly well described with invariant f_is by fixing ΔG_η to the value given in Table 2. It is to be seen from Fig. 7 that the calculated $w_c(t)$-curve displays the sigmoidal bending of the observation. It should be noticed now that crystal growth does not happen

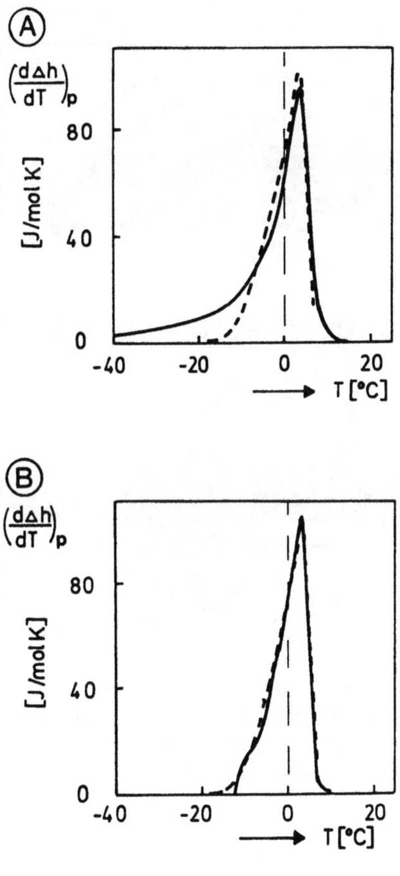

Fig. 5. A) DSC-melting diagrams of natural rubber isothermally crystallized at T = 248 K for 30 h, then heated at a rate of 5 K per minute. The solid lines are the equilibrium melting curves computed by using the parameters collected in Table 1. B) The same measurements as in Fig. 5A described with the aid of the kinetic approach by strictly accounting for the experimental procedures (solid lines, parameters Table 1).

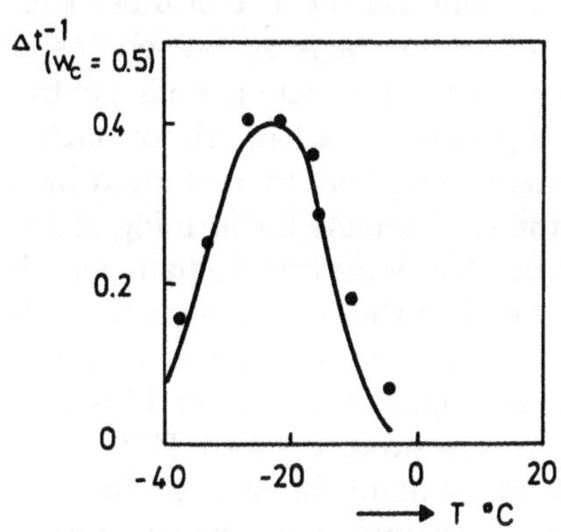

Fig. 6. The reciprocal times for reaching 50% of the equlibrium crystallinity against the crystallization temperature according to [2]. The calculated overall crystallization velocity is depicted by the solid line. Parameters are given in Table 1.

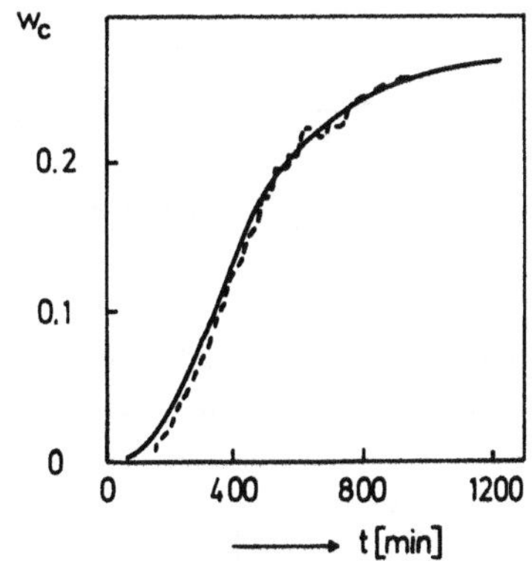

Fig. 7. $w_c(t)$ of natural rubber at T = 248 K. The solid line is computed by using the parameters as depicted in Table 1

to occur in our approach. The sigmoidal bending of $w_c(t)$ is here the consequence of "consecutive and partly simultaneous" nucleation of EMCs of different thickness. During nucleation $\Delta\mu_y$ changes so as to individually alter the degree of undercooling for each fraction of EMCs. The second term in Eq. (46) is related to cooperative effects due to the presence of stable EMCs. This might be considered as a type of secondary nucleation.

The thermodynamically and kinetically equivalent subsystems

Our description of nucleation kinetics in networks is only understandable if the processes develop on very local scales. There should exist thermodynamically and kinetically equivalent subsystems. The same processes run in each of these subsystems. The macroscopic observation is a reflection of local situations. The size of these subsysems must be sufficiently large so that chains of different lengths are available. The low concentration of each component is a significant factor. Within an equivalent subsystem there should be a just sufficient number of chains for forming stable EMCs of each type. The very broad chain length distribution where each of them is linked in the network impedes any crystal growth. This is in accord with the finding that mosaic blocks of EMCs in LDPE have just the size predicted here [41].

It is interesting to estimate the size of the subsystems. Let us now assume that a lower limit of this dimension is given by the mean distance of those chains which participate in forming the largest EMCs (y_{maxim}). A reasonable number is possibly defined by dividing the mean chain length by the mole fraction of the long chains which are built into the first EMCs that can be observed:

$$D^* = \frac{\sqrt{\frac{x_c}{x_{nc}}}}{x_c^{(y_{maxim}-1)}}\ ; \quad y_{maxim} \approx 6\,\frac{x_c}{x_{nc}}\ . \tag{48}$$

$$D = D^* \cdot 0.124 \text{ nm}; \quad y_{max} = y_{maxim} \cdot 0.124 \text{ nm}$$

We learn from Table 3 that the parameter increases with decreasing molar fraction of the *nc*-units. If x_{nc} falls below the value of about 0.02, folded-chain crystallization comes into play. The value of D at x_{nc} = 0.02 may be taken as an upper limit of the mean size of equivalent subsystems. According to Table 3 this maximum size should be in the range of about 300—400 nm. Representative dimensions of superstructure-elements are very often found to lie just in this regime [4, 42].

Strain-induced crystallization in stress-strain cycles

By studying stress-strain cycles under constant strain-rate, one should be able to identify the particular kinetics of strain-induced crystallization or

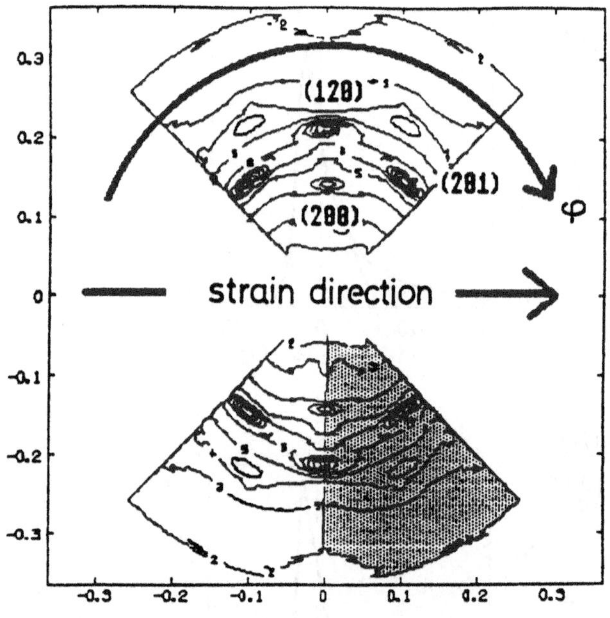

Fig. 8. Iso-intensity contour-plot of a synchrotron pattern from natural rubber strained to λ = 4.5 at room temperature. The Miller indices (hkl) as well as the direction of the extension are indicated with the plot

Table 3. D as the mean size of equivalent subsystems

x_{nc}	0.10	0.075	0.05	0.025	0.02	0.015	0.01	0.001
y_{max}/nm	6.9	9.5	14.5	30.0	37.6	50.4	76.0	767.2
D/nm	100	130	179	284	325	385	482	1586

⇒ chain folding

melting. EMCs nucleated at temperatures above the maximum melting point T_{ymax} are fully oriented (Fig. 8) [1, 2, 13, 40, 43, 44].

I can be seen from Fig. 9 that the typical stress-strain hysteresis of natural rubber is also obtained in the calculation (parameters in Table 2). The theoretical stress-strain curve without crystallization does not deviate very much from the experiment where crystallization happens to occur (thus, some controversis in the literature become understandable [1, 2, 18]).

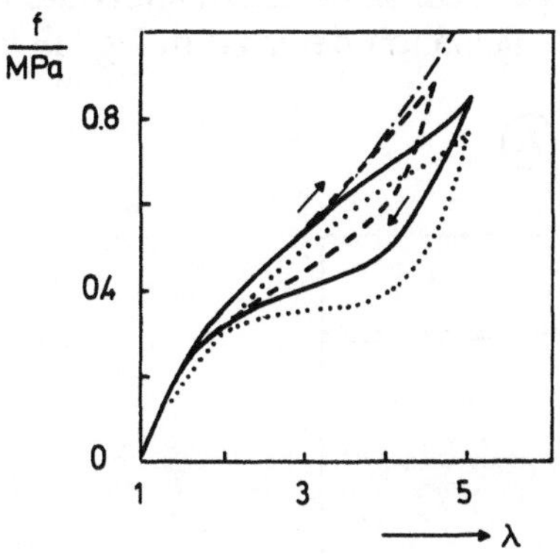

Fig. 9. The dashed line gives the stress-strain cycle of natural rubber at room temperature under the constant strain rate of $\dot{\lambda}$ = 0.03% s^{-1}. The dotted line is from an experiment where the strain is raised stepwise, and after each stretching increment is stopped for 20 min. The solid line is computed for a continuous drawing with the same strain rate as above by using the parameters collected in Table 2. A theoretical stress-strain pattern without crystallization is represented by the dash-dotted line

The calculated curve for the constant strain rate of $d\lambda/dt$ = 0.03% s^{-1} displays a too large hysteresis. But the calculation is not very far from the stress-strain cycle with the strain changed stepwise ($\Delta\lambda$ = .5, after each step holding the strain for 20 min). This corresponds roughly to the same time table as in the constant-strain rate experiment. It is then a proof of consistency that the differential heats, as measured stepwise in a stretching micro-calorimeter, are also fairly well described (Fig. 10).

It is now to be seen that, during the stretch, crystallization starts (as indicated by the downward arrow) while the last EMCs melt at comparably lower strains (upward arrow). Moreover, melting covers that range of strains that is demanded by thermodynamics.

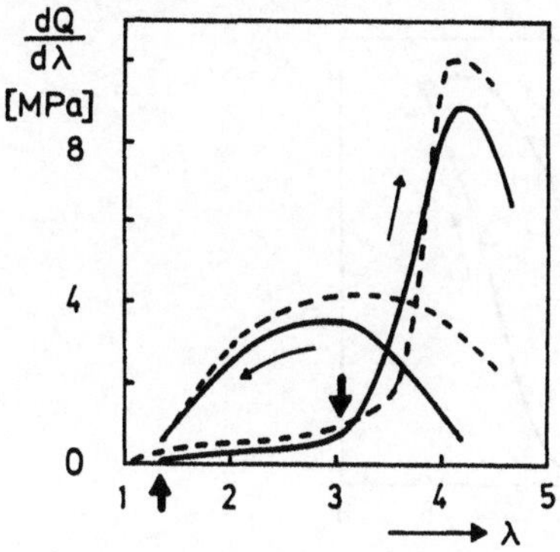

Fig. 10. The differential heats ($dQ/d\lambda$) observed in the stretching micro-calorimeter for a stress-strain cycle by raising the strain stepwise (see capture of Fig. 9). Solid lines are computed by using the same parameters as in Fig. 9

Figure 11 reveals the peculiar hysteresis of the intrinsic strain in the rubbery regions. From Fig. 12, we learn that the maximum crystallinity enforced by straining the network remains remarkably below the maximum value of $w_{c\text{-max}}$ = 0.25.

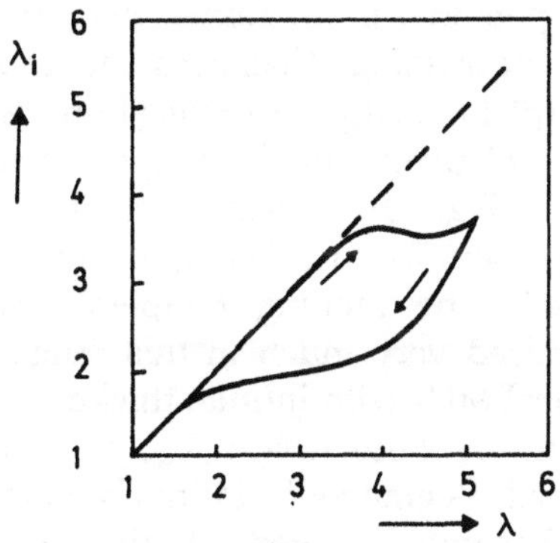

Fig. 11. The intrinsic strain of the natural rubber network against the macroscopic strain during the stress-strain cycle (--- no crystallization)

We come to the following conclusions:

- Fully oriented EMCs formed during strain-induced crystallization seem to have the same intrinsic properties as the ones in a semicrystalline network crystallized from the "melt": This justifies the concept to treat EMCs as thermodynamically autonomous, inhomogeneous microphases.

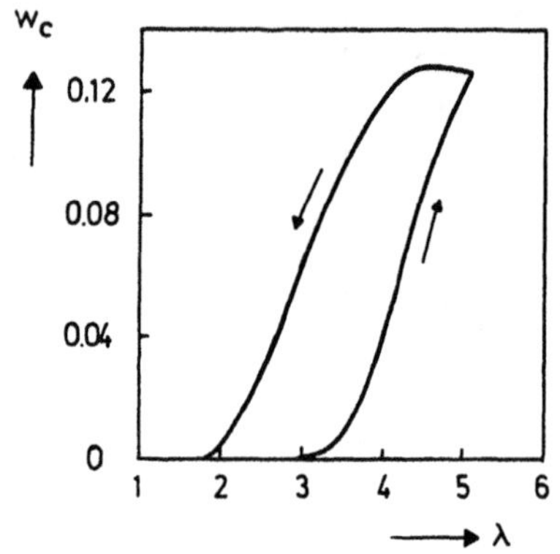

Fig. 12. The molar degree of crystallinity of the natural rubber network during a stress-strain cycle as function of the macroscopic strain

- Strain-induced crystallization starts with largest EMCs followed consecutively by fractions of decreasing thickness. Mainly due to the thermodynamic factor in Eq. (46), EMCs with diferen thicknesses nucleate at different strains. Within a certain regime nucleation of fractions of EMCs with different thicknesses occurs simultaneously. The strain-induced modification in the chemical potential in a strained network increases with decreasing thickness of the EMCs. This explains why crystallization is squeezed into a relatively small strain-interval (upturn of $dQ/d\lambda$ in Fig. 10). For very thick EMCs (which are not present in our system) an inversion is predicted: The melting temperatures should not be raised very much in this range, and go to zero for EMCs with infinite thickness. The melting point of these EMCs, T_M, should not be raised at all. Segments in infinitely long chains cannot become oriented at all. This elucidates the non-Gaussian features of our description.
- Relevant crystal growth seems not to occur.
- The kinetic factor f_2 does not depend on the mode of crystallization (strain- or temperature induced). f_1, on the other hand, is largely increased in the mode of strain-induced crystallization (factor of 100) (see Table 2). The reasons for this characteristic phenomenon are unknown.
- "Overdrawing" is clearly manifested by the reluctant melting just after the strain is released. The course of the differential heats during unloading is fairly well described and reflects the selective and consecutive melting of an eutectoid copolymer. Hence, strain-induced crystallization leads to the same thickness distribution of EMCs as in melt-crystallized samples.
- Internal equilibrium within EMCs, as well as in the rubber regions, is very rapidly established so as to display a reversible thermo-elastic behavior.

Stress-relaxation due to strain-induced crystallization

Stress-relaxation is observed for a rubber stretched to $\lambda = 5$ and kept at constant length afterwards. The quality of the calculation depicted in Fig. 13 is

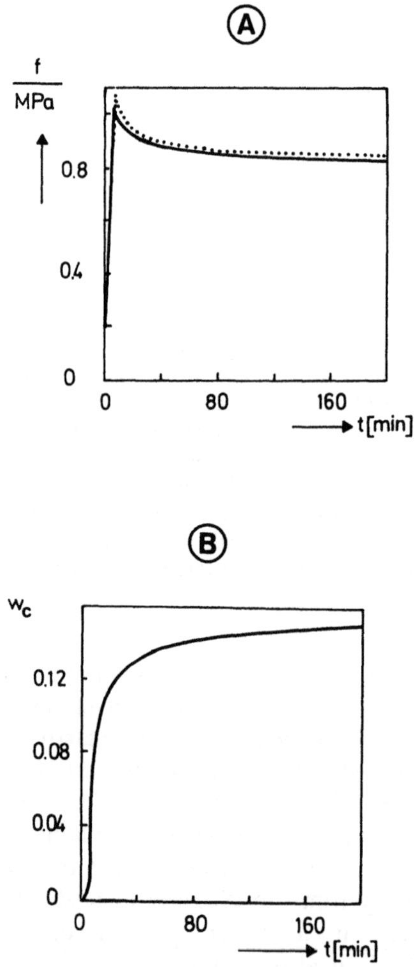

Fig. 13. A) Strain-induced crystallization relaxation of the natural rubber network against the time after being stretched to $\lambda = 5$ with a strain rate of 1% s^{-1}. The dotted line is computed with the parameters collected in Table 2. B) The degree of crystallinity as function of time. It should be noted that the equilibrium degree of crystallinity is far below the maximum value of $w_c = 0.24$

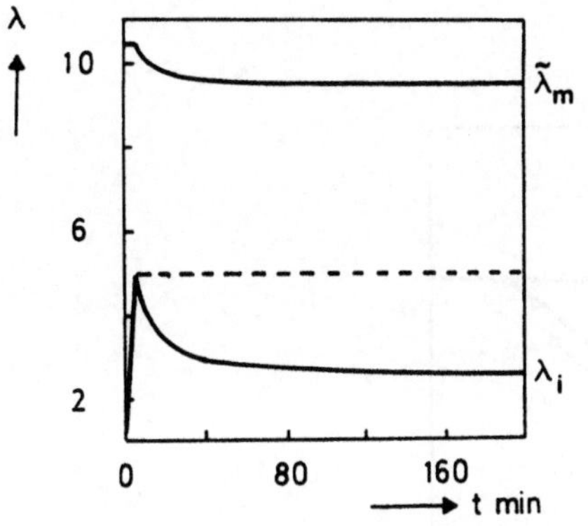

Fig. 14. The same experiment as in Fig. 13: At the top, the maximum strain parameter ($\lambda^2_{max} = y$) in the rubbery matrix during "crystallization-relaxation", as shown in Fig. 13. The plot at the bottom shows the reduction of the intrinsic strain in the rubbery matrix due to continuously forming oriented EMCs

surprisingly good. When the deformation is stopped, stress-relaxation comes about by still forming oriented EMCs. This is supported by the plots in Fig. 14. The configuration of EMCs should correspond to the assumptions which are behind the definition of the intrinsic strain in Eq. (31).

Crystallization or melting in a stretched rubber

In a cooling and heating cycle of a stretched rubber at constant length consecutive nucleation and melting is predicted, yielding to theoretical curves as in Figs. 15–17. But, compared with experiments there are systematic discrepancies (Figs. 15–17). Crystallization starts at temperatures much below the predicted ones (ΔTs up to 30 K). The actual

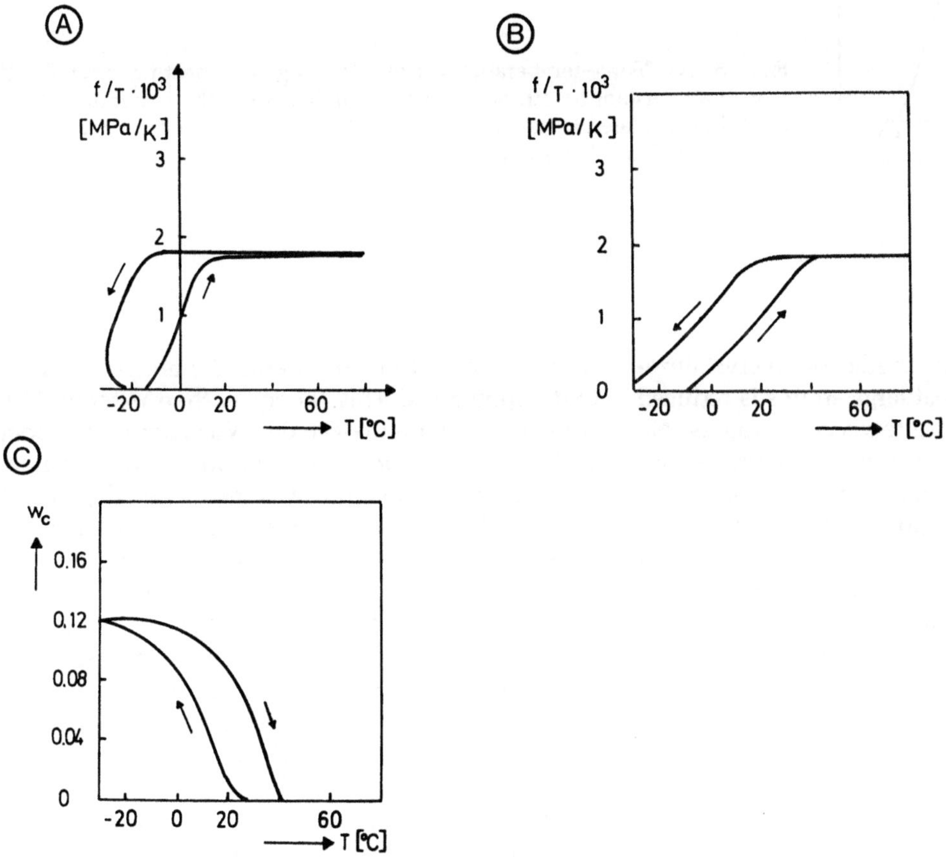

Fig. 15. A) "Force-temperature-cycles" (cooling and heating rate: 0.3 K min^{-1}) at constant length of a stretched rubber to the maximum strain $\lambda = 3$. B) Calculated force-temperature cycles with the aid of the parameters depicted in Table 1. C) Crystallinity

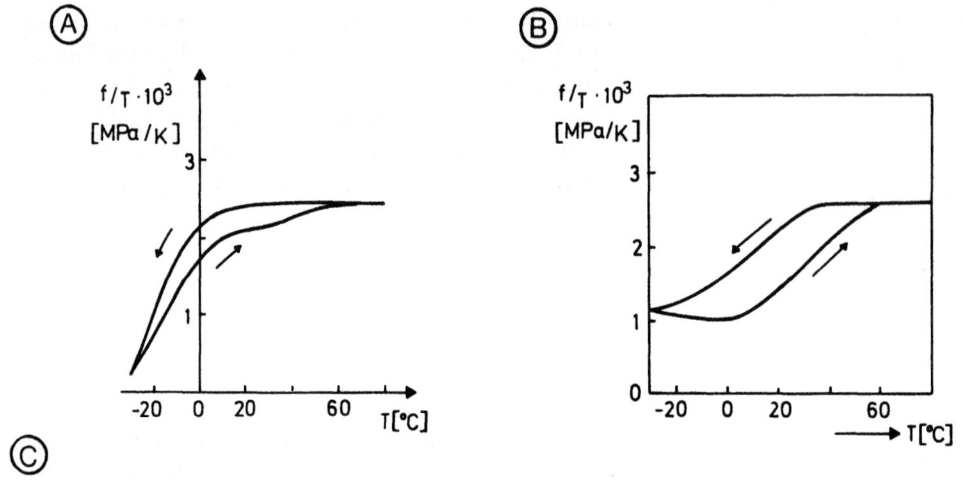

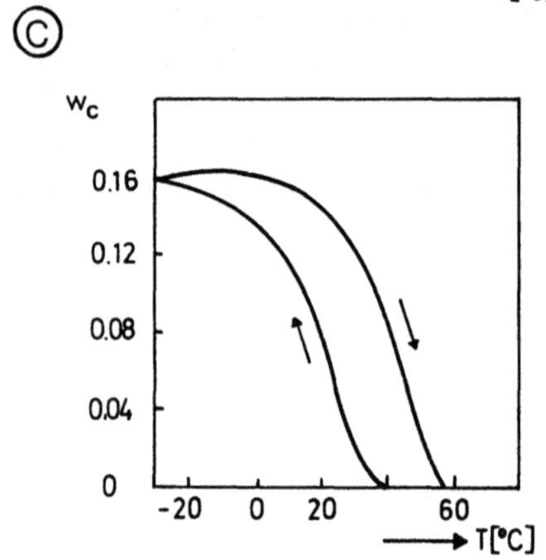

Fig. 16. A) "Force-temperature-cycles" (cooling and heating rate: 0.3 K min^{-1}) at constant length of a stretched rubber up to the maximum strain λ =4. B) Calculated force-temperature cycles with the aid of the parameters depicted in Table 1. C) Crystallinity

nucleation rate exceeds the prediction. If crystallization continues the stress at large strains is reduced at the deepest temperature to very low values. No straightforward interpretation of the discrepancies can be given. ΔG_η might, for example, depend in a strained rubber in a complicated and unknown manner on special situations to which the system is subjected to. It is also conceivable that the configuration of oriented EMCs differs from what is assumed in the mean-field formulation of the intrinsic strain.

Final remarks

We have presented a description of the stress-strain pattern and the total energy balance of thermally and strain-induced crystallization kinetics within the framework of a single theory. This approach is based on the thermodynamics of eutectoid copolymers. This theory is then extended by combination with the van der Waals network model and with a new formulation of nucleation in eutectoid multi-component systems. The degree of undercooling or overdrawing at which nucleation is possible is, in a complicated manner, controlled by thermodynamic factors. When nucleation proceeds the rate is modified because the melting points depend on the continuously changing composition in the multi-component melt. Strain-induced nucleation is, in addition, controlled by the intrinsic strain in the rubbery regions which is reduced when oriented EMCs are formed. The weakest point in our treatment is that the intrinsic strain is defined by disposing of a distinct configuration of EMCs; this might explain the discrepancies encountered. Because all the deviations are systematic, it is hoped that the underlying physical reasons might be identified by means of appropriate experiments.

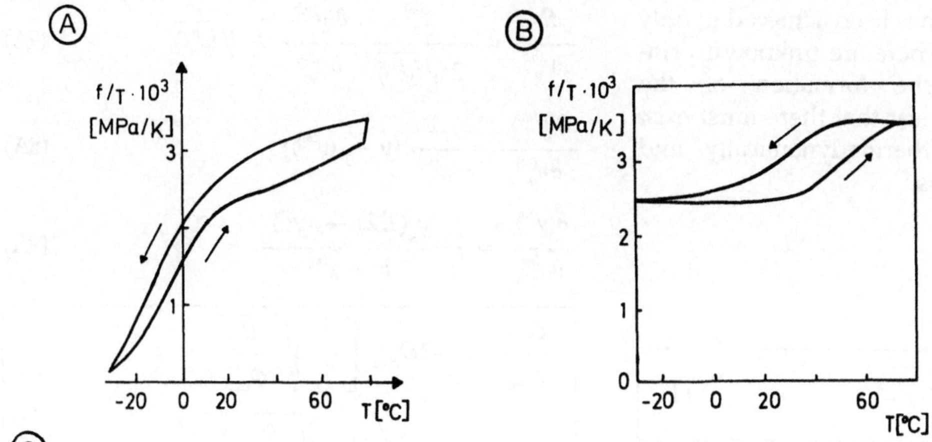

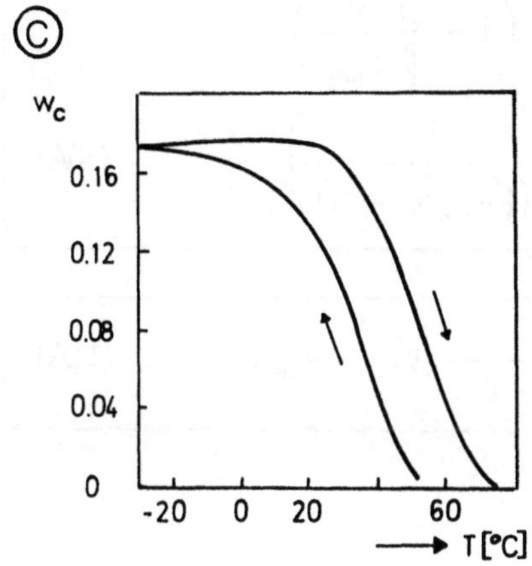

Fig. 17. A) "Force-temperature-cycles" (cooling and heating rate: 0.3 K min^{-1}) at constant length of a stretched rubber up to the maximum strain $\lambda = 5$. B) Calculated force-temperature cycles with the aid of the parameters depicted in Table 1. C) Crystallinity

Crystallization-induced segregation of junctions is evident and always leads to the same EMCs. EMCs are inhomogeneous saturated mixed micro-phases. Moreover, from the description of thermo-elastic phenomena it is evident that internal equilibrium in EMCs and in the rubbery matrix is established. Micro-phases have well defined intrinsic properties. Considering the interfacial contacts, they behave like autonomous subsystems.

Segregation of chains of different lengths must occur on very local scales. The network must be comprised of thermodynamically and kinetically equivalent subsystems, the mean size of which is in the regime of colloidic dimensions. This answers the question of why these multi-component systems come so near to equilibrium. This is supported by necessary adjustments of the composition and the mass fraction of EMCs as demanded for eutectoid systems. Rearrangements of existing EMCs are possible and allow for a continuous optimization of local situations.

The investigation presented here shows some shortcomings. It, nevertheless, supports the view that colloid structure is strictly related to the chemical structure of the network. This most important aspect is introduced by defining stereo-regular chains of different lengths as thermodynamic components. Limited solubility in the solid state is the reason why the primary parameters of the colloid-structure enter into thermodynamics. To define the inhomogeneous mixed microphase is a key assumption.

Crystallization in networks could be interpreted as "self-organization" of a eutectoid multi-component system. But, we hesitate to use this "modern way" of argumentation, because we hope that the presented work identifies the classical roots that lead to a solid understanding of crystallization in networks. The success on the level of the primary structure is encouraging and it is hoped that the reader will understand the senstive rules behind the formation of colloid-structures in eutectoid copolymers. Until now, this has been achieved in only a few exceptional cases. There are unknown principles which control the formation of the superstructure. But, it is clear that there must exist subsystems which are thermodynamically and kinetically equivalent units.

Appendix

$$RT \frac{\partial w_\Sigma}{\partial \lambda_i} = f \tag{1}$$

Using the mass balance $\langle y^{(p)} \rangle$, Eq. (31) can be written as

$$\langle y^{(p)} \rangle = \frac{\langle y_0^{(m)} \rangle - x^{(m)} \langle y^{(m)} \rangle}{1 - x^{(m)}} . \tag{2A}$$

For the same reason, we have

$$\partial n_y^{(c)} = -\partial n_y^{(m)} . \tag{3A}$$

Thus, we are led to the equations

$$\lambda_i = \frac{Z}{N} = \frac{\lambda \sqrt{\langle y_0^{(m)} \rangle} - (1 - x^{(m)}) \left[\left\{ 1 - \frac{A}{3} (1 - \beta) \right\} \langle y^{(p)} \rangle - \left(1 - \frac{A}{3} \right) y_k + \beta \frac{B}{3} \right]}{\sqrt{\langle y^{(m)} \rangle x^{(m)}}} \tag{4A}$$

$$\frac{\partial \lambda_i}{\partial n_y^{(m)}} = \frac{1}{N} \left(\frac{\partial Z}{\partial n_y^{(m)}} - \lambda_i \frac{\partial N}{\partial n_y^{(m)}} \right) \tag{5A}$$

$$\frac{\partial Z}{\partial n_y^{(m)}} = \left(1 - \frac{A}{3} (1 - \beta) \right) \langle y^{(p)} \rangle - \left(1 - \frac{A}{3} \right) y_k + \beta \frac{B}{3} - (1 - x^{(m)}) \left(1 - \frac{a}{3} (1 - \beta) \right) \cdot \frac{\partial \langle y^{(p)} \rangle}{\partial n_y^{(m)}} \tag{6A}$$

$$\frac{\partial N}{\partial n_y^{(m)}} = \frac{x^{(m)}}{2 \sqrt{\langle y^{(m)} \rangle}} \frac{\partial \langle y^{(m)} \rangle}{\partial n_y^{(m)}} + \sqrt{\langle y^{(m)} \rangle} \tag{7A}$$

$$\frac{\partial \langle y^{(m)} \rangle}{\partial n_y^{(m)}} = \frac{1}{x^{(m)}} (y - \langle y^{(m)} \rangle) \tag{8A}$$

$$\frac{\partial \langle y^{(p)} \rangle}{\partial n_y^{(m)}} = - \frac{y_u(T, \lambda) - \langle y^{(p)} \rangle}{1 - x^{(m)}} \tag{9A}$$

$$\frac{\partial w_\Sigma}{\lambda_{max}} = -2 D_{max} \left[\frac{1}{2} \sqrt{\frac{\phi}{\phi_{max}}} + \ln \left(1 - \sqrt{\frac{\phi}{\phi_{max}}} \right) + \frac{1}{2 (\sqrt{\phi_{max} / \phi} - 1)} \right] \tag{10A}$$

$$\frac{\partial \lambda_{max}}{\partial n_y^{(m)}} = \frac{1}{1 - x^{(c)}} [\sqrt{y_\zeta} - \lambda_{max}] \tag{11A}$$

References

1. Holl B, Kilian HG, Schenk H (1990) Coll Polym Sci 268:205
2. Treloar LRG (1958) The Physics of Rubber Elasticity
3. Kim HG, Mandelkern L (1968) J Polym Sci A2, 16:181
4. Wunderlich B (1976) Macromolecular Physics, Vol 2, Academic Press, New York, San Francisco, London
5. Müller FH, Engelter A (1957) Koll Z Z Polymere 152:15
6. Engelter A, Müller FH (1958) Koll Z Z Polymere 157:89
7. Göritz D, Müller FH (1973) Koll Z z Polymere 251:892
8. Kilian HG, Höhne G, Trögele P, Ambacher H (1984) J Polym Sci Symp 77:221
9. Göritz D (1982) Coll Polym Sci 26:193
10. Godovsky YK (1986) Adv Polym Sci 259:75
11. Heise B, Kilian HG, Schmidt H (1981) Colloid Polym Sci 259:611
12. Kilian HG, Unseld K, Jaeger E, Müller J, Jungnickel B (1985) Colloid Polym Sci 263:607
13. Holl B, Kilian HG, Yeh GSY (1985) Colloid Polym Sci 263:313
14. Kilian HG (1988) Progr Colloid Polym Sci 78:161
15. Asbach GI, Kilian HG (1991) Polymer 32:16
16. Glenz W, Kilian HG, Klattenhoff D, Stracke F (1977) Polymer 18:685
17. Kuhn W, Grün F (1942) Koll Z 101:242
18. Mark JE, Erman B (1988) Rubberlike Elasticity — A Molecular Primer, Wiley, New York, Chichester, Brisbane, Toronto, Singapore
19. Kilian HG (1981) Polymer 22:209

20. Kilian HG (1989) Makromol Chem, Macromol Suppl 30:169
21. Vilgis T, Kilian HG (1986) Colloid Polym Sci 264:131
22. Vilgis T, Kilian HG (1986) Colloid Polym Sci 264:137
23. Kilian HG, Enderle HF, unseld K (1987) Progr Colloid Polym Sci 75:55
24. Zrinyi M, Kilian HG, Horkay F (1989) Colloid Polym Sci 267:311
25. Paul E, Heise B, Schrodi W, Kilian HG (1991) Progr Colloid Polym Sci 85:12
26. Kilian HG, Enderle HF, Unseld K (1986) Colloid Polym Sci 264:866
27. Ambacher H, Strauß M, Kilian HG, Wolff S (1991) Kautschuk + Gummis. Kunststoffe 44:1111
28. Wood LA (1946) Advances in Polymer Science, II, Interscience Publishers, New York
29. Göritz D, Müller FH, Sietz W (1977) Progr Colloid PolymSci 62:114
30. Mitchell JC, Meyer DJ (1968) J Polym Sci A-2, 6:1689
31. Yamamoto M, White J (1971) J Polym Sci A2, 9:1399
32. Oth JF, Flory PJ (1958) J Am Chem Soc 80:1297
33. Mukherji S, Woodward AC (1984) J Polym Sci, Phys E 22:793
34. Pei-guo Wang, Woodward AC (1987) Macromolecules 20:1818
35. Gavish M, Corrigan J, Woodward AC (1988) Macromolecules 21:2079
36. Allegra G, Bruzzone M (1983) Macromolecules 16:1167
37. Holl B, Kilian HG, Yeh GSY (1985) Colloid Polym Sci 263:313
38. Siesler H, Perl W (1988) Microchim Acta I:323
39. Flory PJ (1947) J Chem Phys 15:397
40. Mandelkern L (1964) Crystallization of Polymers, Mc Graw, New York
41. Asbach GI, Bodor G, Kilian HG (1989) Colloid Polym Sci 267:976
42. Heilemann P (1992) Diplom-Thesis, Abteilung Experimentelle Physik, University of Ulm
43. Luch D, Yeh GSY (1972) J Appl Phys 43:4326
44. Luch D, Yeh GSY (1973) J Macromol Sci Phys B7:121

Authors' address:

Prof. Dr. H. G. Kilian
Abt. Experimentelle Physik I
Universität Ulm
Oberer Eselsberg
7900 Ulm, FRG

Complete balance of energy during necking in polymeric glasses and its strain-rate dependence

J. A. Koenen and H.-G. Kilian

Universität Ulm, Abt. Experimentelle Physik, FRG

Abstract: Some polymeric glasses, e.g., polycarbonate, deform heterogeneously after yielding by forming a neck. The boundary between two zones with different strains λ' and λ'' propagates at constant stress with a valocity v_n over the sample, changing the deformation of the material from λ' to λ''. This process is accompanied by an exchange of mechanical work and heat. At low v_n both quantities are measured in a stretching calorimeter. At higher v_n, we observe the sample in the stretching device with an infrared camera. Comparison of the measured temperture pattern on the surface of the sample with a calculation using the differential equation of heat conduction including heat sources and heat loss results in the strength of the sources and, therefore, to the heat produced by the transition. The mechanical work shows a small increase with v_n. The heat is constant at low v_n and increases at $v_n \approx 2 \cdot 10^{-4}$ m/s until the entire work is transformed into heat.

Key words: Deformation; polycarbonate; necking; energy balance, strain-rate dependence; infrared camera

The mechanical properties of polymeric glasses are a field of major research [1—3]. Some materials deform heterogeneously after yielding, e.g., polycarbonate, PMMA. Further stretching is done by propagation of the boundary between the two zones with different strains λ', λ'', called the neck [4]. A more complete understanding of this phenomenon can be achieved by investigation of the mechanical and the thermodynamical properties of necking [5, 6, 7]. The thermomechanical behavior, investigated with a stretching calorimeter [8], shows clearly that the necking process displays the characteristics of a phase transition from a low strained state λ' to a high strained state λ''. The extension of such measurement to higher strain rates is possible for mechanical tests, but not for the stretching calorimeter. Therefore, we use an infrared camera and observe the temperature image on the surface of the sample [9]. The comparison of this temperature pattern with calculations using the equation of heat conduction including heat sources and heat loss leads to the strength of the sources and, therefore, to the heat produced. Together with the mechanical data measured simultaneously, we obtain the complete balance of energy.

For all experiments we use polycarbonate samples (Macrolon 2800) with length 100 mm, width 10 mm, thickness 4 mm, and dogbone shaped (DIN 53 455 No. 3). The experiments are performed in a Zwick 1445 tensile tester. The local strains are obtained by measurements of a line pattern on the surface of the sample. Therefore, we observe the experiment with two video cameras, one shooting the sample, the other the force display, and the IR camera. A clock and title generator adds information about the experiment and time.

The IR part of the film is transformed from a thermal to a temperature image using the calibration of the IR camera and the emissivity of the sample. All experimental data (stress strain curve, local strains, λ', λ'', neck velocity, temperature profiles, etc.) are evaluated from the film using the image-processing system (Imaging Technology PFG+) plugged into a PC.

In the model given in [7], we assume that the force is a combination of two parts. One part accounts for the chain structure of the polymeric glass and the strain energy is set proportional to the first strain invariant. The second part has its origin in the solid state properties of the glass. Analysis of the model yields to limits of stability and a phase transition at a constant force f_t from $\lambda' \rightarrow \lambda''$.

In the region of coexistence the whole deformation takes place at the boundary, called the neck, between the phases. Therefore, the heat of this transition, as measured with a stretching calorimeter, is produced exactly at the position of the neck. This heat should be produced homogeneously over the cross-section of the sample and will disperse according to the laws of heat conduction; it will be lost to the surrounding due to heat transfer, and will be transported with the deformation of the material. The one-dimensional equation of heat conduction is written as [10]

$$\frac{\partial^2}{\partial x^2}\vartheta - \frac{1}{\kappa}\frac{\partial}{\partial t}\vartheta = -\frac{A}{K} + \frac{\nu}{\kappa}\vartheta \,. \quad (1)$$

The necking process is responsible for the heat sources A, and we assume an amount of heat q_t per volume being produced. Therefore, we get

$$A(x,t) = q_t\delta(x - v_n t) \,. \quad (2)$$

The last term in Eq. (1) describes the heat loss over the length of the sample, which is proportional to the temperature difference ϑ between sample and surrounding, and has been determined by a Newtonian cooling experiment.

The solution of this differential equation, after changing to the comoving frame of reference ($x' = x - v_n t$) yields

$$\vartheta(x') = \frac{q_t v_n \sqrt{\kappa}}{2\sqrt{\pi}\,K}\int_{-\infty}^{\infty} d\tau\,\Theta(\tau)e^{-\nu\tau}\frac{1}{\sqrt{\tau}}\,e^{-\frac{(x'+v_n\tau)^2}{4\kappa\tau}} \,. \quad (3)$$

Taking into account the convection of heat due to materials deformation a comparison of the measured temperature profiles and the calculations with Eq. (3) yields the power of the heat sources q_t and, together with results from the tensile experiment, yields the complete energy balance of the necking process in dependence of neck velocity v_n.

Because strain rate makes no sense when deformation becomes heterogenous, the neck velocity v_n is used to characterize the speed of deformation. Besides this, strain rate is a function of machine speed and sample length, while the neck velocity is a function of machine speed only.

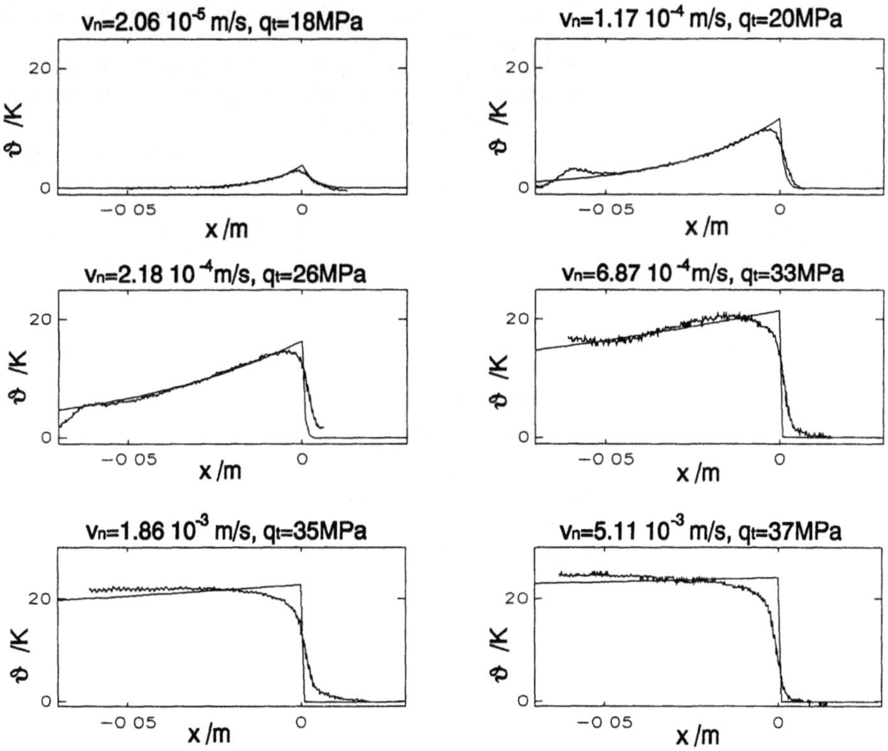

Fig. 1. Measured and calculated temperature profiles of polycarbonate at room temperature with different stretching rates (m: meter)

The measured temperature profiles parallel to the stretching direction at different neck velocities are shown in Fig. 1. We can see a transition from nearly isothermal to adiabatic deformation. The descriptions of the measured temperature profiles with Eq. (3) are also included in the picture.

To check the correctness of our method, we compare the results at low neck velocities with values measured in a stretching calorimeter [7]. In both cases, we get 18 MPa (= mJ/mm^3) heat per volume for the transition $\lambda' \rightarrow \lambda''$. This exchange is independent of neck velocity at low speeds and is exactly the change of orientational entropy as measured with birefringence. Frozen in stresses yield to an increase of internal energy.

The plot in Fig. 2 shows the mechanical work w_t and the heat q_t in dependence of the neck velocity. At the highest strain rates the entire mechanical work is transformed into heat. Because it is not reasonable to produce more heat than work done (this may be possible only in processes like strain-induced crystallization or similar, which are not observed in PC [9]) we arrive at a second fixpoint of the complete energy balance. Therefore, it is shown that the method with the IR camera measures the entire heat produced.

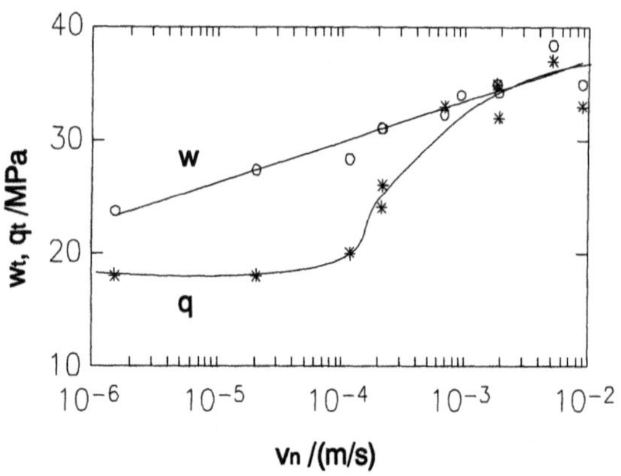

Fig. 2. Work and heat of the transition from $\lambda' \rightarrow \lambda''$ in dependence of the neck velocity

At a neck velocity of about $2 \cdot 10^{-4}$ m/s, we see a strong increase of q_t from a constant value at lower v_n until it reaches the work w_t. This behavior may be a relaxational effect or a cooperative process, transforming the entire work into heat at high stretching rates due to reduction of the frozen in stresses. Therefore, at low neck velocities internal energy increases with the transition $\lambda' \rightarrow \lambda''$, while at the highest velocities it remains constant.

The method developed to measure the complete energy balance during tensile tests is an extension of the measurements in a stretching calorimeter and may be denoted as "fast calorimetry".

References

1. Haward RN (1973) The Physics of Glassy Polymers, Applied Science Publishers, London
2. Kovacs AJ, Aklonis JJ, Hutchinson JM, Ramos AR (1979) J Polym Sci: Polym Phys Ed 17:1097
3. Theodorou DN, Suter UW (1986) Makromolecules 19:379
4. Hutchinson JW, Neale KW (1983) J Mech Phys Solid 31:405
5. Müller FH (1949) Kolloid Z 114:59
6. Adams GW, Farris RJ (1988) J Polym Sci: Polym Phys Ed 26:433
7. Koenen JA, Kilian HG, Heise B (1989) J Polym Sci: B: Polym Phys 27:1235
8. Müller FH, Engelter A (1957) Kolloid Z 152:15
9. Maher JW, Haward RN, Hay JN (1980) J Polym Sci: Polym Phys Ed 18:2169
10. Carslaw HS, Jaeger JC (1959) Conduction of Heat in Solids, Clarendon Press

Authors' address:

Dr. Joachim A. Koenen
Universität Ulm
Abteilung Exprimentelle Physik
Albert-Einstein-Allee 11
W-7900 Ulm, FRG

Progress in Colloid & Polymer Science Progr Colloid Polym Sci 87:107—109 (1992)

Morphology and structure of oriented polydiacetylene films: Molecular epitaxy of DCH

A. Thierry*), F. Kajzar°), and J. Le Moigne*)

*) I.C.S. 6, rue Boussingault Strasbourg, France
°) CEA Saclay Gif sur Yvette, France

Abstract: The morphology and structure of oriented polydiacetylene films have been studied by optical and electron microscopy. The thin films are prepared under vacuum by epitaxial nucleation of the monomer crystal on single crystal substrates. Later on, a thermal topochemical polymerization preserves the orientation of the molecules and, hence, of the polymer chains. We thus obtain different types of organization of the chains depending on the interactions with the monocrystalline substrate. The films are polycrystalline. The higher nucleation density on organic substrates gives rise to a fine film texture. Third-order nonlinear characteristics have been measured; an anisotropy of 140 has been observed.

Key words: Epitaxial growth; single orientation; texture; thin films; polydiacetylenes

Introduction

Conventional synthetic macromolecules usually crystallize in chain-folded lamellae, forming either spherolites for bulk crystallization or tiny single crystals of a few µm in size in the case of solution growth. Special crystallization conditions are required to obtain fully extended chains. On the contrary, polydiacetylenes (pDA) naturally form macroscopic single crystals a few mm in size with fully extended chains. This peculiar organization was first reported by Wegner [1] in the case of the poly(hexadiynetoluenesulfonate). It is attainable through a solid-state topochemical polymerization of the parent monomer. However, some crystallogenesis problems still exist when one wants to prepare materials for integrated optics. Growth of large, defect-free, single crystals of monomer is difficult and the subsequent solid-state polymerization introduces new defects owing to the crystal lattice modifications [2].

Nevertheless, if charge-transfer complexes are suitable for second-order nonlinear effects, π-conjugated polymers are appropriate for third-order effects; the second hyperpolarizability (γ) being proportional to the fifth power of the conjugated double bonds. PDA polymers have exceptionally high susceptibilities and exhibit a subpicosecond response off-resonance [3].

Therefore, highly oriented thin films of π-conjugated polymers are of practical importance. We have taken advantage of the high order of organization attainable through epitaxial growth to enhance their macroscopic third-order susecptibility, $\chi(3)$, and prepare thin films. This method, currently used to grow layers of inorganic materials, has not yet been much favored for organic ones. Other lining-up techniques have already been used for diacetylene chains [4—6]; the advantage of epitaxy is that it is very efficient and versatile for all polymers, even intractable ones.

The polymer we selected for this work was poly(1,6-di(N-carbazolyl)-2,4 hexadiyne) (pDCH) (see monomer below). The degree of polymerization is one of the highest that can be reached for diacetylenes ($\geqslant$98%), and the polymer itself is highly resistant to oxidation and thermal degradation [6].

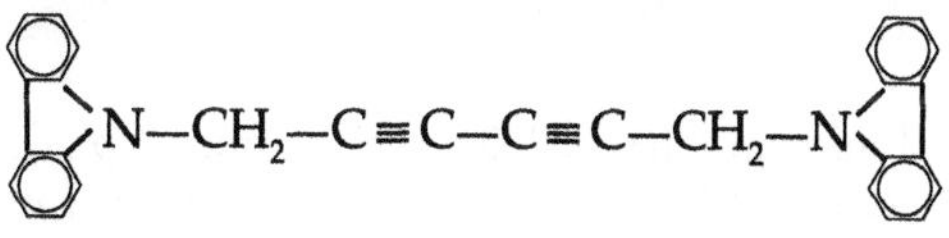

Experimental section

Experimental conditions have already been described in detail elsewhere [8]. Thin films of monomer are prepared by rapid sublimation under high vacuum (10^{-5} Pa). The vapors are condensed on different crystalline substrates: (i.e., potassium bromide, phthalic-acid alkali salts, 2—5 piperazinedione, α-phenylglycine) and the thickness of the film is monitored by a quartz balance. The substrates are obtained with two main methods: fresh cleavage of a monocrystal (KBr) or solution evaporation for organic crystals. Part of the monomer is lost by thermal polymerization in the crucible. The purity of the film and the polymerization rate can be checked by visible linear spectroscopy. We prepared monomer films 20 to 250 nm thick. The subsequent thermal solid state polymerization is performed at 150 °C for 24 h.

The morphology and structure of these oriented films are characterized by optical and electron microscopy. Optical microscopy is performed on a K Zeiss photomicroscope II, electron microscopy is performed on a Hitachi HU 11 CS or a Philips CM 12. Their measured nonlinear characteristics are described elsewhere [9].

Morphology and structure of the films

The morphology and, hence, the optical qualities of the films vary with the substrate. They are always polycrystalline and show different types of organization. There are two orientations at right angles to each other on cubic substrates (KBr), two orientations also on 2—5 piperazinedione, but at 18° to each other, and a single one each on phthalic-acid potassium salts and on α-phenylglycine.

Pleochroism of the mono-oriented films is illustrated in Fig. 1. When the chains are parallel to the polarization the absorption is maximum (blue film); when they are perpendicular to it the film is transparent.

The crystallites' size also depends on the substrate nature. Two populations of rather rough crystals are present on mineral subtrates [8], whereas for 2—5 piperazinedione and for α-phenylglycine a single population of small crystallites (~20 nm width ~ 400 nm length) is nucleated (Figs. 2A, 3A). A higher degree of nucleation on organic substrates seems to be a general rule for the substrates tested so far. The quality of the orientation is checked with electron diffraction. The diffraction patterns are formed of slightly arced spots (angular dispersion of ~6°); two examples are shown in Figs. 2B, 3B. The contact plane observed has always been ($\bar{1}02$). The small arced spots prove the uniformity of the mono-orientation.

Comments

Detailed studies of nonlinear optic measurements on these films have already been published [8, 10]. In the present work we emphasize that the observed light-scattering intensity is weak and that the nonlinear anisotropy measured on mono-oriented samples can be as high as 140.

We clearly demonstrate that variable morphologies, some of them very promising for practical developments, can be observed on organic films epitaxially grown. This should, of course, apply for other kinds of organic materials with specific properties (i.e., second-order nonlinearity, ferroelectric or conductivity properties).

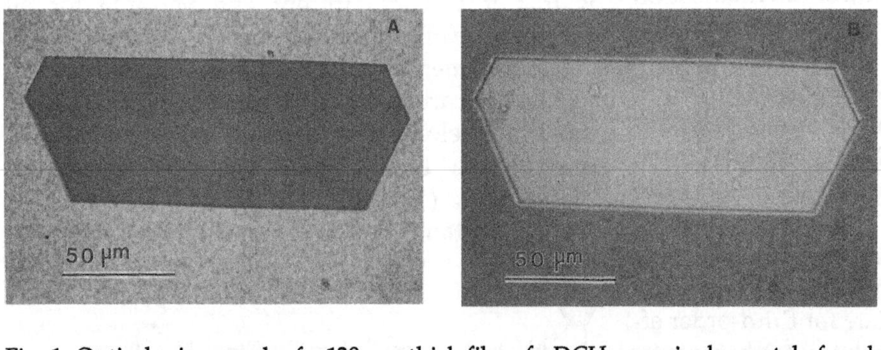

Fig. 1. Optical micrograph of a 130-nm-thick film of pDCH on a single crystal of α-phenylglycine observed in polarized light. Polarization parallel (A), and perpendicular (B) to the polymer chains

Fig. 2. Electron micrograph of a 42-nm-thick film of pDCH on 2—5 piperazinedione: A) bright field; B) diffraction pattern

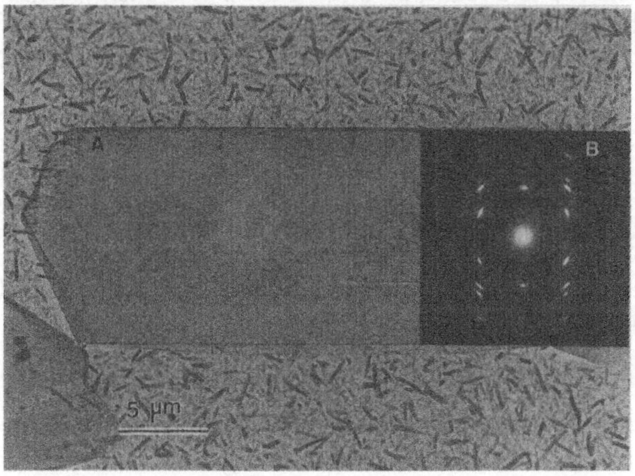

Fig. 3. Electron micrograph of a 20-nm-thick film of pDCH on α-phenylglycine: A) bright field; B) diffraction pattern

The size of the films can be controlled by the size of the single crystals used as substrate, and they can be quite large (see [10, 11]). We confirmed in a previous work that the epitaxial orientation is kept throughout the film thickness (see [8]).

References

1. Wegner G (1969) Z Naturforsch 24B:824
2. Young RJ, Read R, Petermann J (1981) J Mater Sci 16:1835
3. Carter GM, Thakur MK, Chen YJ, Hrynievicz JV (1985) Appl Phys Lett 47:457
4. Patel JS, Sin-Doo L, Baker GL, Shelburne JA III (1980) Appl Phys Lett 56:131
5. Takur M, Meyler S (1985) Macromolecules 18:2341
6. Kanetake T, Ishikawa K, Hasegawa T, Koda T, Takeda K, Hasegawa M, Kubodera K, Kobayashi H (1989) Appl Phys Lett 54:2287
7. Batchelder DN, Poole NJ, Bloor D (1981) Chem Phys Lett 81:560
8. Le Moigne J, Thierry A, Chollet PA, Kajzar F, Messier J (1988) J Chem Phys 88(10):6647
9. Yee KC, Chance RR (1978) J Polym Sci, Polym Phys Ed 16:431
10. Le Moigne J, Kajzar F, Thierry A (1991) Macromolecules 24:2622
11. Le Moigne J, Thierry A, Guerder P (1989) Eur Pat 89 403 536.9

Authors' address:

A. Thierry
Institut Charles Sadron
6, rue Boussingault
67083 Strasbourg Cedex, France

Morphology and extensibility of dried gels from an ethylene copolymer in relation to initial concentration

O. Darras, R.Seguela, and F. Rietsch

Laboratoire des Structures et Propriétés de l'Etat Solide, URA CNRS 234,
Université de Lille, Villeneuve d'Ascq, France

Abstract: Dried gels of a linear low-density ethylene copolymer cast from decalin solutions are investigated from the standpoint of structural, thermal, and mechanical properties. In the concentration range 1.00—0.20 the swollen gels exhibit a nearly isotropic shrinkage upon drying which is the expected behavior of an ideal crosslinked network. For concentrations below the critical value $C^{\dagger}$ = 0.20, non-isotropic shrinkage is obtained, indicating that the chains begin to disengage from the macromolecular network due to the dilution. The melting behavior of the dried gels shows that crystallization from solution notably improves the perfection of the more defective crystals. The concomitant decrease of the crystal thickness judged from small-angle x-ray scattering is ascribed to a reduction of the surface free energy, which is consistent with the build-up of regular chain-folded macroconformations. The extensibility of the dried gels is considerably improved with increasing dilution as a result of the gradual disentanglement of the coils prior to the crystallization in solution. But at concentrations below $C^{\dagger}$ = 0.20, the extensibility drops because of the loss of intermolecular cohesion when the chains begin to disengage from the network. The fact that $C^{\dagger}$ is much greater than the critical concentration for gelation shows that a large molecular overlap is necessary for chain intertwining.

Key words: Ethylene copolymer; dried gels; entanglements; thermal behavior; tensile drawing

Introduction

We recently suggested that the correlation between the drawing behavior and the crystallinity of linear ethylene copolymers relies on a change of the chain topology as a function of the co-unit concentration because of the co-unit exclusion from the crystal lattice [1, 2]. Indeed, the crytallization of high-density copolymers is likely to take place through a regular chain-folding mechanism with a concomitant reeling effect and a disentanglement of the coiled chains from the melt (intramolecular crystallization), whereas a solidification process without long-range rearrangement of the chains should occur in the case of co-unit rich, low-density copolymers (intermolecular crystallization).

Solution crystallization was particularly well suited to confirm the above conclusions because of the dilution effect which promotes regular chain-folded macroconformations together with chain disentanglement. This has been previously proposed by several authors in the case of ultrahigh-molecular-weight polyethylenes which exhibit a considerably improved extensibility when crystallized from dilute solutions [3—7].

Experiments

The material studied is an ethylene/1-butene copolymer containing about 8 mole% of co-units. Its number- and weight-average molecular weights are M_n = 32 000 and M_w = 160 000 and its nominal density is ρ = 0.910 g cm^{-3}. The gels were prepared into flat discs from decalin solutions by means of a mold fitted with an injection piston held at the temperature of 150 °C. The mold cavity had a thickness of 2 mm and a diameter of 140 mm. The crystallization-induced gelation was performed by plunging the mold into cold water. The gels were dried under vacuum for 2 days, and subsequently washed in a methanol bath for 2 days.

The differential scanning calorimetry (DSC) measurements were performed on a Perkin Elmer DSC-7 Delta at the heating rate of 10 °C/min. The tensile drawing experiments were carried out with an Instron Tester at a cross-head speed of 50 mm/min using 25 mm gauge length samples. At least three measurements have been performed for each sample, and the scatter on the values of both the yield stress and ultimate draw ratio did not exceed 5%. The small-angle x-ray-scattering (SAXS) patterns have been recorded with a Luzzati-Baro camera, and the long periods have been computed from the maximum of the scattering peak by applying the Bragg's relation.

The critical concentration for gelation has been determined by the ball drop method of Takahashi et al. [8], using 2 mg steel balls, as suggested by Okabe et al. [9]. The absolute accuracy was about 2.5×10^{-3}.

The molecular weight distribution and the average molecular weights have been obtained from gel permeation chromatography (GPC) in trichlorobenzene at 140 °C.

Results and discussion

The plot of Fig. 1 shows that the diameter-to-thickness ratio of the dry gel discs is very close to that of the mold cavity in the concentration range 1.0—0.2. This means that the shrinkage of the discs upon drying is isotropic and gives evidence that the swollen gels behave like ideal polymer networks. But below the critical concentration $C^† = 0.2$ the disc thickness is much more reduced than the diameter, indicating that the process of solvent removal is accompanied by a sedimentation effect which involves a staking of the lamellar crystals under the force of gravity.

The DSC melting curves of the dried gels reported in Fig. 2 show an improved crystal perfection with increasing dilution, notably as concerns the more defective crystals. Indeed, a high-temperature melting peak grows at the expense of the low-temperature melting tail, and the crystallinity is slightly improved. The concomitant decrease of crystal thickness judged from SAXS measurements (Table 1) is relevant for a reduction of the surface free energy, which is consistent with the build-up of more regular chain-folded macroconformations.

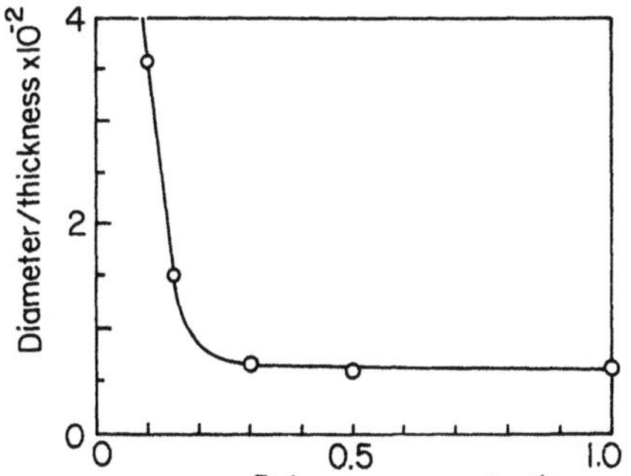

Fig. 1. Plot of the diameter-to-thickness ratio of the dried gel discs as a function of the concentration of the initial solutions

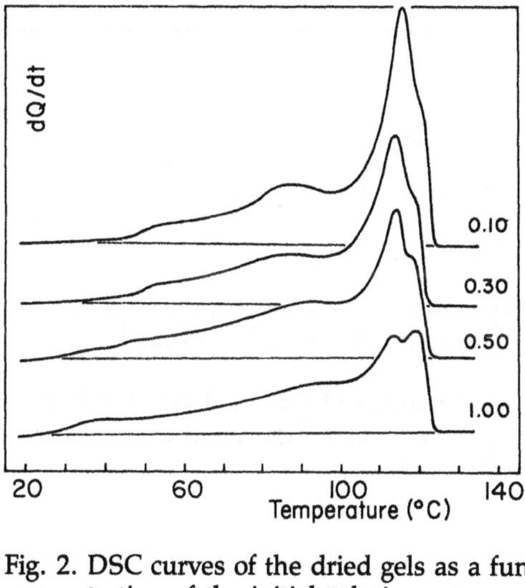

Fig. 2. DSC curves of the dried gels as a function of the concentration of the initial solutions

Table 1. Crystal thickness, *Lc*, and crystal weight fraction *), *Xc*, as a function of initial concentration, *C*

$C(^w/_w)$	1.00	0.90	0.50	0.30	0.10
Lc(nm)	diffuse halo	9.6	8.0	6.2	5.8
Xc	0.35	0.34	0.37	0.39	0.45

*) From DSC assuming $\Delta H_f^0 = 290$ J/g.

The extensibility of the dried gels is strongly improved by the effect of the increasing dilution up to the concentration C = 0.3, as can be seen from the drawing curves of Fig. 3. Considering that the maximum draw ratio of a polymer network is related to the number of random links between entanglements by

$$\lambda_{max} = n^{1/2} ,$$

with the Graessley relation [10]

$$(M_e)_{solution} \approx (M_e)_{melt}/\Phi ,$$

where M_e is the molecular weight between entanglements and Φ is the volume concentration of the solution, the above result is a clear indication of a reduction of the entanglement density of the chains in the solution prior to crystallization. The decrease of the strain-hardening effect in the drawing curves is in agreement with a more slack macromolecular network, by analogy with the rubber elasticity theory which predicts a softening of

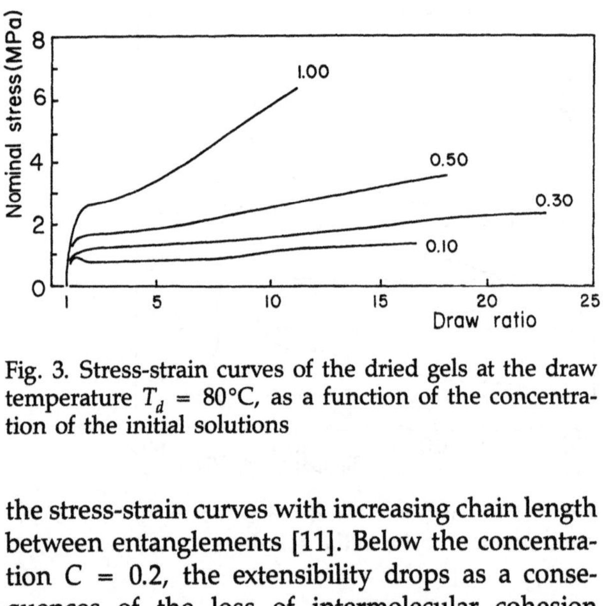

Fig. 3. Stress-strain curves of the dried gels at the draw temperature T_d = 80°C, as a function of the concentration of the initial solutions

the stress-strain curves with increasing chain length between entanglements [11]. Below the concentration C = 0.2, the extensibility drops as a consequences of the loss of intermolecular cohesion when the chains begin to disengage from the network. Indeed, considering that $(M_e)_{melt} \approx 2000$ in the case of polyethylene (12), it follows that $(M_e)_{solution} \approx 10\,000$ at $C^\dagger = 0.20$. Then, because $M_n = 32\,000$, this means that the chains contain only two entanglements on average. The molecular weight distribution curve of Fig. 4 shows that, in fact, about 25% of the chains are such that $M < M_n$ and, therefore, cannot contribute to the physical network. The proportion of inactive chains then becomes catastrophic for $C < C^\dagger$. This conclusion is perfectly consistent with the divergence from isotropic shrinkage of the gels upon drying and supports the assignment of $C^\dagger = 0.20$ as the critical concentration for an efficient chain intertwining.

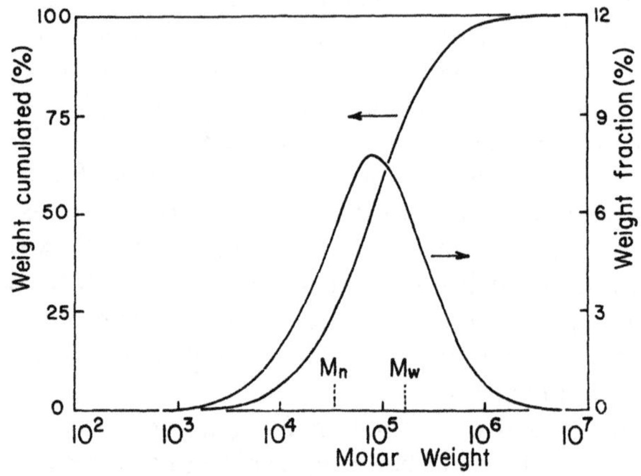

Fig. 4. Molecular weight distribution curve of the copolymer from GPC

This parameter could be compared to the critical concentration, C_{ec}, for the occurrence of chain entanglements in the core regions of the coils as defined by Aharoni [13]. $C^\dagger$ is much greater than the critical concentration for gelation, $C^* \approx 0.03$, providing evidence that a large molecular overlap is necessary for chain intertwining. Indeed, it is obvious that the coils have to interpenetrate deeply into each other in order that both ends of every chain could diffuse by reptation through the neighbor chains and experience intercrossing events with them.

References

1. Seguela R, Rietsch F (1986) Polymer 27:703–708
2. Seguela R, Rietsch F (1988) J Mater Sci 23:415–421
3. Kanamoto T, Sherman ES, Porter RS (1979) Polym J 11:497–502
4. Smith P, Lemstra PJ, Booij HC (1981) J Polym Sci, Polym Phys Ed 19:877–888
5. Pennings AJ, Smook J (1984) J Mater Sci 19:3443–3450
6. Matsuo M, St. John Manley R (1982) Macromolecules 15:985–987

7. Lemstra PJ, Kirschbaum R (1985) Polymer 26:1372—1384
8. Takahashi A, Sakai M, Kato T (1980) Polym J 12:335—341
9. Okabe M, Mitsui K, Sasai F, Matsuda H (1989) Polym J 21:313—328
10. Graessley WW (1974) Adv Polym Sci 16:1—179
11. Ward IM (1985) Mechanical Properties of Solid Polymers, 2nd Ed, Wiley Interscience, New York, pp 62—78
12. Porter RS, Johnson JF (1966) Chem Rev 66:1—27
13. Aharoni SM (1978) J Macromol Sci-Phys B15:347—370

Authors' address:

Dr. Roland Seguela
Laboratoire des Structures et Propriétés de l'Etat Solide
Bât. C6
Université de Lille I
59655 Villeneuve d'Ascq Cedex, France

Processing effects

Polymer solidification by crystallization under heat transfer and flow conditions*)

H. Janeschitz-Kriegl

Inst. of Chemistry, Johannes-Kepler-University Linz, Austria

Abstract: Two limiting cases can be discriminated when solidification processes are considered in which crystallization is involved. With relatively thick samples and high crystallization speeds the process is heat-diffusion controlled, whereas with thin samples and low crystallization speeds the process becomes nucleation-rate controlled. Solidification processes in crystallizable polymers clearly demonstrate the latter type of process. Surprisingly, no mathematical tools had been developed so far for the correct description of the local interplay between heat flow and evolution of heat by crystallization, as controlled by the crystallization kinetics. A review is given of recent developments in this field, which is complicated by the changes in crystallization kinetics, as introduced by the action of shear flow.

Key words: Polymers; flow; heat transfer; crystallization; DSC

Introduction

Solidification processes by crystallization can be investigated in quiescent melts which do not remember any previous mechanical treatment, or in melts influenced by previous or continuing flow. As is well-known, stirring has been known to have a tremendous influence on crystallization kinetics. Its influence forms a special field of research in industrial crystallization from solutions. With polymer melts, a related influence is exerted by the mold-filling procedure and by the extraordinary memory of polymers to previous flow. But, first, we have to consider the more simple situation of relaxed quiescent polymer melts. We do this in following a generally accepted concept, viz. to work through a complicated matter step by step. Surprisingly, however, even with quiescent polymer melts, no previous systematic research exists. Such a study was carried out recently in the course of a national working party in Austria. A concise review is attempted here. For more information the final report by this study group [1] and a recently published review [2] should be consulted. Obviously, only an amalgamation of principles as taken from transport phenomena (heat transfer) and from crystallization kinetics can be successful.

With solidification in quiescent melts, one has already to discern between two important limiting cases [3, 4]:

a) *Heat-diffusion controlled processes,* occurring in thick samples and/or with rapidly crystallizing materials. Molding of parts made from metals or their alloys can often be characterized by this type of process.

b) *Nucleation rate-controlled processes,* occurring in thin-walled samples and/or with slowly crystallizing materials. Most processing situations of polymer melts are clearly illustrative for this case. Of course, only thin-walled, sometimes even extremely thin-walled parts are produced by injection molding or by other techniques used in polymer processing.

Thus, the treatment of polymers seems particularly simple, since for case b) the coupling between

---

*) Dedicated to Prof. O. Vogl, Herman F. Mark Professor, Polytechnic University Brooklyn, on the occassion of his 65th birthday.

Plenary lecture at EPS 23, Stockholm, June 1991

heat transfer and crystallization kinetics is particularly weak. There is even an additional advantage: the heat of crystallization of polymers is much smaller than that of metals. Therefore, the heat-source term in the equation of heat conduction is small, even if the speed of crystallization is not very low. As a consequence, one can solve the equation of heat conduction first, ignoring crystallization (and consequent changes of heat diffusivity and density). In a second procedure, the kinetic equations can be integrated over the local thermal history as obtained in the first procedure. In most cases, an iteration appears unnecessary. As we will see, there is only one exception: full coupling must be considered for differential scanning calorimetry (DSC), because of the fact that at thermal signal is used. This is a kind of an irony, since almost everyone has advocated this technique for the evaluation of crystallization kinetics of polymers. Later, we will address this in more detail.

General principles for quiescent melts

Kolmogoff's equation

In 1937, Kolmogóroff [5] formulated the principles of nucleation and growth of crystals. For spherulitic growth, as characteristic for quiescent polymer melts, he arrived at the following equations in an exact way from statistical considerations:

$$\xi(t) = 1 - \exp\left\{-\frac{4\pi}{3}\int_0^t ds\,a(s)\left[\int_s^t dz\,G(z)\right]^3\right\} \quad (1)$$

and

$$N_c(\infty) = \int_0^\infty ds\,a(s)[1 - \xi(s)]\,, \quad (2)$$

where ξ is the fraction of space covered by spherulites at time t, $a(t)$ is the rate of nucleation at time t, $G(t)$ is the linear growth speed of the spherulites at that time, and $N_c(\infty)$ is the ultimate number of spherulites per unit of volume (at $t = \infty$). The introduced time dependences are of a quite general nature. A special case is that a and G are unique functions of temperature T, the latter being a function of time according to the cooling process considered. (With induction by flow, one will discover a second reason for a time dependence.)

The difficulty with these equations is that they are of the integral type. In fact, with cooling processes they are coupled to the equation of heat conduction, which is a partial differential equation. The latter equation describes the local temperature changes, as influenced by the local progress in crystallization:

$$\rho c\frac{\partial T}{\partial t} = k\Delta T + \rho h\frac{\partial \xi}{\partial t}\,, \quad (3)$$

where ρ is the density, c the heat capacity, and k the heat conductivity which are assumed to show no considerable temperature dependence, whereas h is the effective heat of crystallization which is the heat of crystallization for 100% crystallinity multiplied by the degree of crystallinity of the newly formed spherulites. (Very often this quantity can be conveniently obtained from the surface area of the DSC-peak.)

The rate equations

A regular interplay according to Eq. (3) can only be effect with the aid of a differential version of Eq. (1). (We are grateful to Schneider, Köppl, and Berger [6], members of our Austrian working party, for introducing their "rate equations".) For this purpose, the logarithmic version of Eq. (1) was differentiated several times with respect to time. At each step an auxiliary function φ_i (with $i = 1, 2, 3$) was introcued. In this way, one obtains the following set of differential equations:

$$\varphi_i(t) = \frac{1}{G(t)}\frac{d}{dt}\varphi_{i-1}(t)\,, \quad (4)$$

with initial conditions

$$\varphi_i(0) = 0 \text{ for } i = 0, 1, 2, 3. \quad (4a)$$

Explicitly, one has

$$\varphi_0 = -\ln[1 - \xi(t)]\,, \quad (1a)$$

and

$$\varphi_3 = 8\pi\int_0^t ds\,a(s) \quad \text{or} \quad \dot{\varphi}_3 = 8\pi a(t)\,. \quad (5)$$

The second Eq. (5) forms the starting point for the integration. With respect to the texture, one has the additional equation:

$$\dot{N}_c(t) = a(t)\exp[-\varphi_0(t)]\,. \quad (6)$$

The "rate equations" are completely equivalent to Eqs. (1) and (2). The physical meaning of the auxiliary functions is (except for numerical factors):

φ_0: unrestricted spherulite volume, ignoring impingement, per unit of volume;
φ_1: projection of this volume on a unit surface;
φ_2: its projection on a unit line (edge of unit cube);
φ_3: unrestricted number of activated nuclei, per unit of volume, disregarding the effect of swallowing, as caused by spherulites which started growing at an earlier time.

The present formulation was proposed by Eder [7] on the basis of Kolmogóroff's equations. The authors of the rate equations started from Avrami's equation, and arrived at a slightly different formulation.

Avrami's model

In 1939 and 1940, Avrami [8] treated the following special case, as characterized by the following parameters:

$n = n(T)$: frequency of activation of potential nuclei, as a function of temperature;
$\bar{N}$: number of potential nucleation sites per unit of volume, being constant in as far as volume changes by temperature are disregarded. (Assumption of a simple heterogeneous nucleation mechanism.)

If $n(T)$ is replaced by $1/\tau(T)$, with τ being a temperature-dependent activation time, one arrives for the auxiliary function φ_3 at:

$$\varphi_3(t) = 8\pi\bar{N}\left[1 - \exp\left(-\int_0^t ds/\tau(s)\right)\right] . \quad (7)$$

This equation shows that, after some time, φ_3 becomes a constant. This means that all potential sites are activated. If this happens in isothermal crystallization before the growth of these nuclei becomes important (ξ is still practically zero), the ultimate number of spherulites is equal to $\bar{N}$ (see Eq. (2)). At the same time, φ_0 (as appears in the exponent of Eq. (1)) becomes proportional to t^3 (Avrami exponent 3!). If the conditions for crystallization change to a lower crystallization temperature or even to anisothermal cooling conditions, one can never get a larger number of spherulites than $\bar{N}$. Now, we know from experience that, for many industrial polymers, Avrami exponent 3 is found for only slight supercooling. These results were achieved with the aid of dilatometric measurements [9—11], which are accurate only at low levels of supercooling where crystallization happens slowly enough. Thus, the conclusion must be, for those polymers, that Avrami's model is unable to predict the tremendous increase of the ultimate number of spherulites experienced with increasing degree of supercooling or with more rapid quenching (which results in crystallization at lower temperatures). Therefore, this model is not generally suitable for predicting one of the most important experimental facts which is linked to product quality: many small spherulites mean many tie-molecules connecting these spherulites and causing good mechanical properties [12, 13].

Similar unsuitable results are obtained with Nakamura's [14] "isokinetic" approach. That author assumes $G\tau$ = const (in our notation). In other words, kinetics of nucleation and growth should show the same temperature dependence. With such an approach any influence of the thermal history on the texture (number of spherulites per unit of volume) is erased per definition [15]. Apparently, such an assumption is not realistic.

Introduction of a distribution of activation mechanisms

The only possible conclusion is that crystallization kinetics are more complex. The proposal has been made to introduce a distribution or a "spectrum" of activation mechanisms [2]. In principle, such a spectrum can easily be introduced into the equations presented. However, a practical evaluation of a large set of parameters $\bar{N}_j$, $\tau_j(T)$, becomes virtually impossible. Therefore, we have restricted ourselves to a simplified spectrum in which, during cooling, each activation time jumps from infinite to zero at the pertinent "temperature of activation". In this way, one gets

$$\varphi_3(t) = 8\pi \sum_{T_j > T(t)} \bar{N}_j = 8\pi N(T(t)) . \quad (8)$$

One only obtains a number of activated nuclei (per unit of volume) which increases continuously with decreasing temperature. This is congruent with an earlier proposal by Van Krevelen [16].

Evaluation of parameters for quiescent melts

Linear growth speed of spherulites

The linear growth speed of spherulites as a function of temperature was considered previously in a rather large number of experimental and theoretical papers. Works by Mandelkern [11], Magill [17–19], and Van Krevelen [9, 10, 16] should be cited in this connection. For industrial polypropylenes Krobath [20] collected data from direct observation of spherulite growth and from zone crystallization, and complemented these data by results obtained with the aid of transcrystallization against a steep temperature gradient in suppressing sperulitic growth in the bulk. For a temperature range bounded by a temperature $T_k = (T_m + T_g)/2$ halfway between glass transition and melting point and by the melting point T_m he found a very reasonable approximation:

$$G(T) = G_r e^{-\beta(T-T_r)} , \tag{9}$$

with $\beta = 0.916\ K^{-1}$ and $G_r = 1.259 \cdot 10^{-6}$ m/s for $T_r = 115\,°C$.

Number of activated nuclei. Microscopy combined with DSC

On the other hand, reliable data for $N(T)$ are scarce. Eder [21], inspired by work by Billon and Haudin [22], and motivated by our varying successes by our ups and downs with DSC-measurements, proposed to investigate the cross-section through samples solidified in the DSC-pan during cooling runs at various cooling rates.

In Figs. 1 and 2 such cross-sections are shown for an industrial polypropylene (Daplen KS 10), as obtained after cooling runs at cooling rates of 5 K/min and 50 K/min, respectively [21]. Pertinent peak temperatures of the DSC-curves were found at 117° and 103 °C, respectively. First, one notices that the size of the spherulites is much smaller with the second run, which means that many more spherulites are formed at the lower crystallization temperature around 103 °C. Second, one notices that also the surfaces of the samples which were in contact with the aluminum walls of the pan (bottom and cover) must have been active in nucleation. If the nuclei on the surface are closer to each other than in the bulk, one obtains a surface layer. The phenomenon has been called transcrystallization. Apparently, the crystallization nucleated in the bulk competed with the crystallization nucleated at the walls.

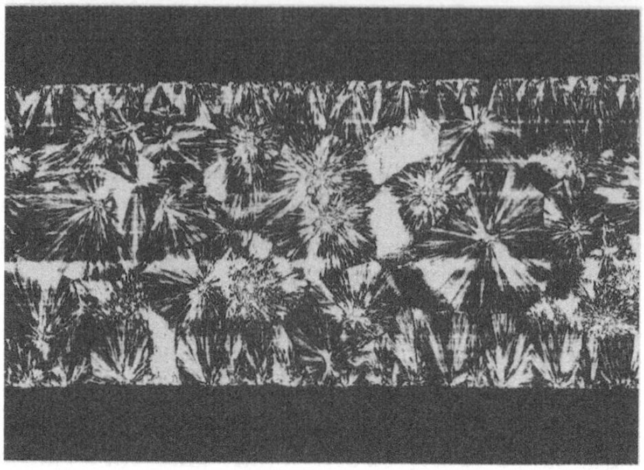

Fig. 1. Cross-section through a sample of industrial polypropylene Daplen KS 10, as obtained from the DSC-pan after a cooling run of 5 K/min, according to [21]

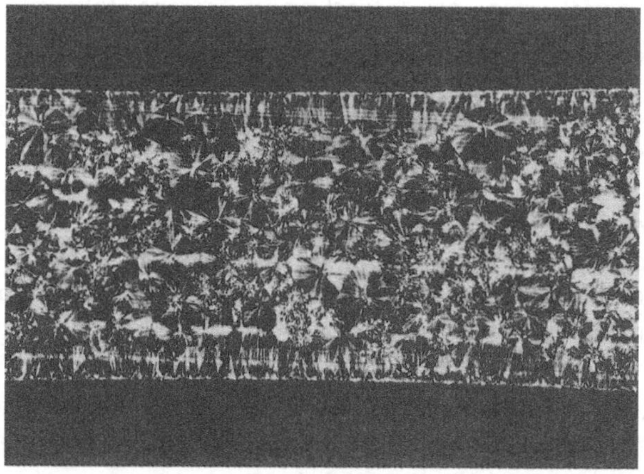

Fig. 2. As in Fig. 1, after a cooling run of 50 K/min ([21])

As Eder has shown [21], one can deduce the number of nuclei per unit of volume in the bulk from the thickness of the transcrystallized layer. (The number of nuclei per unit surface at the wall becomes irrelevant if there are enough of them.) The crystallization starting at the wall runs against the spherulites nucleated in the bulk. Indeed, as one notices from the figures, the surface layers have an irregular boundary against the bulk. The best way of evaluation is to draw an ink line on the photograph and determine the area between this line and the straight line indicating the surface by planimetry. This area, divided by the length of the said straight line, gives the average thickness of the transcrystallized layer, if a sufficient length is chosen.

A simplified version of Eder's explanation is given here. For this purpose, the assumption is made that the crystallization occurs practically at the peak temperature of the DSC-curve. It is also assumed, in accordance with our simplified version of the distribution of activation mechanisms, that all nuclei characteristic for this temperature, are activated instantaneously. In this way, one arrives at the following equation for the degree of space filling by newly formed spherulites:

$$\xi = 1 - \exp\left[-\frac{4\pi}{3} N G^3 t^3\right] , \tag{10}$$

which is Avrami's equation with exponent three (for isothermal crystallization). At the same time, one obtains for the thickness of the transcrystallized layer:

$$\delta = G t . \tag{11}$$

In accordance with the idea that a plane at the average distance δ from the surface cuts through 50% transcrystallized and 50% spherulitic matter, the critical value for ξ is chosen to be 0.5.

From Eqs. (10) and (11) the product Gt is eliminated. One obtains:

$$N(T) = \frac{0.166}{\delta^3(T)} . \tag{12}$$

(In this way, also the influence of $G(T)$ is eliminated.) In Fig. 3 the number of nuclei as a function of temperature is plotted for PP Daplen KS 10, as obtained by Paulik with this method. In this figure the peak temperatures were corrected for the influence of the finite heat flow resistance between pan and furnace of the DSC-apparatus. In this way, the real crystallization temperatures in the samples are found to be higher by some degrees, dependent on the cooling speed. But the general picture, as conveyed by Fig. 3, is not seriously altered by these corrections. As one can see from the figure, one obtains for $N(T)$ an approximate presentation similar to that for $G(T)$:

$$N(T) = N_r e^{-\bar{\beta}(T-T_r)} , \tag{13}$$

holding in the same temperature range. For the parameters, one has, as can be seen from Fig. 3: $\bar{\beta} = 0.18\ K^{-1}$ and $N_r = 2 \cdot 10^{11}\ m^{-3}$ at $T_r = 115\,°C$.

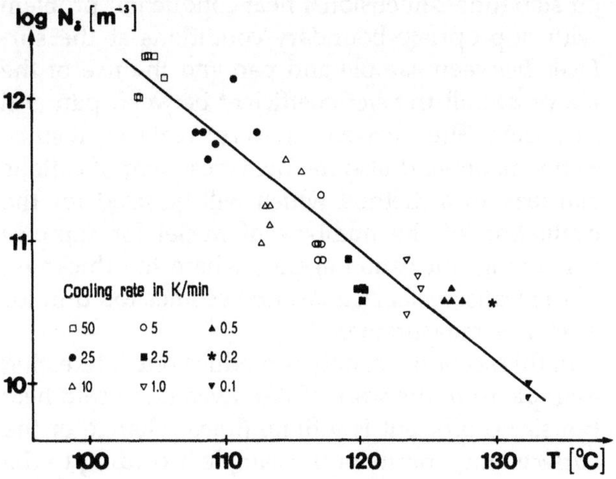

Fig. 3. Numbers of nuclei per unit of volume against peak temperatures of the corresponding DSC-cooling runs, according to [21]

Direct evaluation of the number of activated nuclei from DSC

The heat transfer between pan and furnace, and also within the sample (of relatively low heat conductivity) is of great influence on the results of an evaluation of $N(T)$, if the shape and the position of the DSC-curve are to be used for this evaluation without any additional information as, for example, that obtained from the texture. If, for our crystallization model, heat transfer problems are ignored (the temperature in the sample being assumed to be identical with the furnace temperature) a change in the cooling speed only causes a shift of the position of the peak maximum on the temperature scale. The shape of the peaks remains unchanged. Such a prediction does not seem in accordance with the general experience, according to which the peaks not only shift to lower temperatures with increasing cooling rates, but also become lower and broader. For our crystallization model, we have been able to show that this is a consequence of the finite heat flow resistance between sample, pan and furnace. As a first step, we measured the heat transfer coefficient between pan and furnace with the aid of calibration metals like indium (or tin and lead). For our configuration (a power compensating Perkin Elmer, Model No. 7, with a certain type of pan), we found a value of 16 mW/K [23]. With this avlue, Wu of our laboratory carried out a full-scale simulation [24], considering the sample as an infinitely extended slab (one-dimensional heat conduction problem with appropriate boundary conditions at the surfaces between sample and pan and the use of the above overall transfer coefficient between pan and furnace). She arrived at very realistic results. Refinements and also mathematical simplifications can lead to a method which will be used for the evaluation of the numbers of nuclei for samples containing nucleation agents, where the thickness of the transcrystallized layer δ becomes too thin for a reliable measurement.*)

In this section, we only reproduce one interesting example from the work of Wu. Even if the said heat transfer coefficient is infinite (linear change of the surface temperature of the sample according to the program of the DSC-apparatus), the internal heat flow resistance remains of influence. This is shown in Fig. 4 for our polypropylene. The sample thickness was assumed to be 0.9 mm. For a thinner sample the shapes of the peaks obtained for the cooling rates, as given in the caption to the figure, become more similar to the one at the lowest cooling rate and shift closer to each other and to higher temperatures. In the limit of zero sample thickness, one obtains the picture for which heat transfer problems are ignored (the peaks become equal in shape, but still show a spread in their positions).

This is the first time that such effects can be demonstrated. In fact, without a kinetic model for the crystallization of the polymer, one is unable to predict those phenomena. It should be said, however, that other kinetic models may be found in the future which will predict peak broadening on their own, if the heat transfer problems are ignored. But this possibility does not invalidate our warning with respect to the enormous influence of the finite heat flow resistance. A mere calibration of the temperature scale of the DSC-apparatus for different cooling and heating rates does not correctly solve this problem. In fact, every calibration metal has its own kinetics, different for cooling and heating (crystallization and melting). Even, if this calibration could be done correctly, many problems would remain: among others, one obtains an extra delay for the peak maximum with increasing thickness of the polymer sample because of the low heat conductivity of this material.

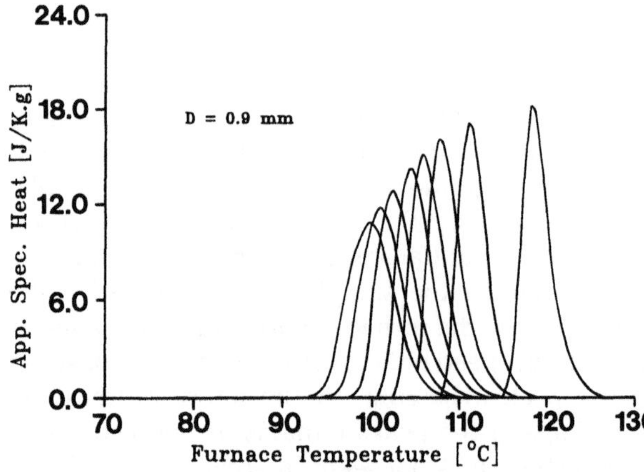

Fig. 4. DSC-curves as simulated for polypropylene with the aid of our crystallization model for a finite sample thickness of 0.9 mm, external heat flow resistance being assumed to be zero (ideal temperature history of DSC-apparatus), according to [24]. Cooling rates: 50, 40, 30, 20, 15, 10, 5, and 1 K/min

*) The reader is reminded that the knowledge of $G(t)$ is needed here, in contrast to Eder's method of the previous section 3.2.

Flow-induced crystallization

Quenching experiments

The melt of a crystallizable polymer must undergo some mechanical treatment, before a kind of internal metamorphosis can become evident. This internal change appears to occur also at a surprisingly high temperature at which its consequences cannot be observed. Under such a condition only a subsequent quench is able to produce a crystallization pattern as normally linked with flow-induced nucleation and growth. If this quench is delayed, the internal change apparently relaxes [2, 20, 25]. The oriented surface layer shown in Fig. 5 was obtained in an extrudated band of polypropylene: after extrusion through a lengthy duct of rectangular cross-section of large aspect ratio (1 × 10 mm^2) this band was quenched on a chill-roll. The extrusion temperature was as high as 200 °C. This surface layer shows a lamellar texture, as found in injection-molded parts [1, 26].

Similar surface layers are also formed in a duct of rectangular cross-section if, at elevated teperature,

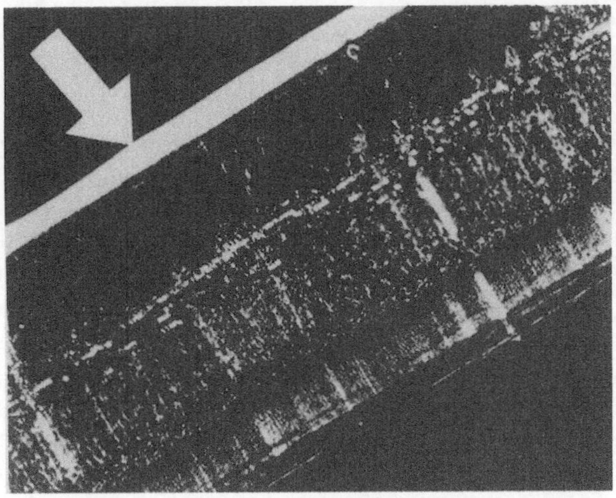

Fig. 5. Highly oriented surface layer in polypropylene, as obtained by a quench after shearing at elevated temperature (200 °C), according to [25] (courtesy of Dr. Alfred Hüthig Verlag, Heidelberg)

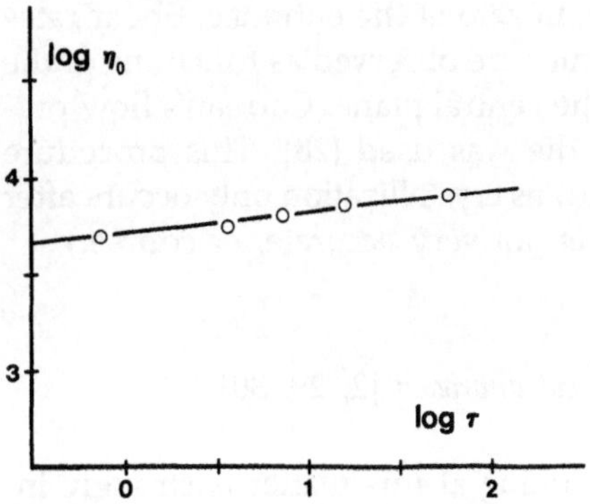

Fig. 6. Double-logarithmic plot of zero shear viscosity of polypropylene Daplen KS 10 against relaxation time of the ability to form oriented structures after the quench. From left to right the temperature of flow varied from 21° to 190 °C. (From [2], courtesy of Pergamon Press)

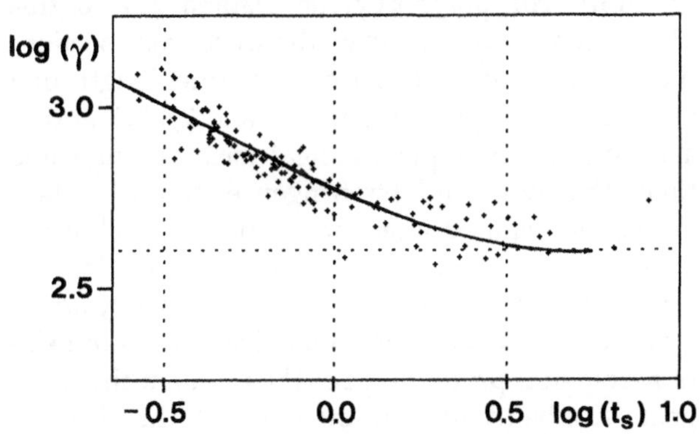

Fig. 7. Double-logarithmic plot of critical shear rate vs. shearing time for polypropylene Daplen KS 10 for a shearing temperature of 200 °C. Below the (scatter of) points no lamellar crystallization was found after the quench (cf. [2])

flow is stopped and the duct is quenched as a whole. With polypropylene the content of the duct can easily be extracted after such a quench. The cross-section can be investigated under the microscope. In particular, the birefingence of the surface layers, as occurs at the large side walls of the duct, can be measured in this way. This birefringence appears to decrease if, with no change in the conditions of previous flow, the time interval between cessation of the flow and the quench is increased. From this decrease of birefringence ($\log \Delta n$ decreases linearly with t) a relaxation time characteristic for the temperature of flow can be deduced [2, 27]. In Fig. 6 the zero shear viscosity of the investigated industrial polypropylene is plotted against this relaxation time for various flow temperatures. From left to right, this temperature varied with the experimental points from 210° to 190 °C. Apparently, this relaxation time does not obey the normal time-temperature superposition which can be expected for the zero shear viscosity. Otherwise, in this double logarithmic plot a line under 45° would be obtained. The conclusion is that a supermolecular structure, as formed by flow, is relaxing. In an interval of 20 °C, from 210° to 190 °C, this relaxation time increases from about 1 s to about 60 s!

Another result of quenching experiments [2, 27] is shown in Fig. 7. Here, the logarithm of the shear rate is plotted against the logarithm of the shearing time for the same polypropylene treated at 200 °C. In the space below the scatter of points no lamellar crystallization occurs after an immediate quench. The line through the points is drawn according to a model discussed below. An important feature is the slope of −1/2 at short shearing times and high shear rates. At large shearing times a lower limit is shown for the shear rate, at which lamellar crystallization occurs. In the logarithmic plot shown this fact is not particularly convincing. However, on a linear plot this is seen very clearly. (These results are obtained from cross-sections of samples extracted from a duct. In these cross-sections the thickness of the highly oriented layer increases continuously from zero at the entrance. Shear rates and shearing times are observed as functions of the distance from the central plane. Carreau's flow profile for the slit die was used [28]. This procedure seems permitted, as crystallization only occurs after the quench. It is not very accurate, of course.)

Tentative theoretical approach [2, 29, 30]

Our way of looking at this matter is strongly influenced by the observation that also the build-up of those hidden structures (precursors) takes time. The establishment of steady-state flow is definitely achieved earlier (cf. the discussion of Fig. 6). Here, we only mention the results of flow birefringence measurements, as carried out near the ends of ducts of widely varying lengths (from 12 mm to 117 mm). For polyporpylene Daplen KS 10 this birefringence (average over the duct cross-section) did not show any influence of the duct length in a temperature range between 180° and 220°C, and in a wide range of apparent shear rates. At the same time, thickness and birefringence of the surface layers, as obtained after the quench on a chill-roll tremendously increased with duct length. Therefore, the reader should not immediately adopt a formulation which does not reflect the complexity of rheological properties. For the moment the main goal must be to find a frame in which the observations readily fit.

In Fig. 7, we note the slope of —1/2 for short shearing times. It seems to be a reasonable assumption that some internal parameter Φ of the melt must reach the same critical value Φ_{crit} along the curve through the points of Fig. 7. For the region with slope —1/2, one obviously has for this value

$$\Phi_{crit} = \left(\frac{\dot{\gamma}}{\dot{\gamma}_a}\right)^2 \frac{t_s}{\tau}, \tag{14}$$

where $\dot{\gamma}$ is rate of shear and t_s is the shearing time. In keeping Φ dimensionless (it may be a kind of probability function), one formally has $\dot{\gamma}_a$ as a shear rate of activation and τ as a relaxation time. From the figure, one can also observe that no shear-induced crystallization occurs at shear rates $\dot{\gamma} < 400\ s^{-1}$.

The most simple differential equation for Φ, which has a solution satisfying these two observations, reads:

$$\dot{\Phi} = \frac{\dot{\gamma}^2}{\dot{\gamma}_a^2 \tau}(1 - \Phi) - \frac{\Phi}{\tau}, \tag{15}$$

where the first term on the right side is a growth term and the second one is a decay term.

A parameter presentation of the isothermal solution of Eq. (15) ($\dot{\gamma}_a$ and τ being kept constant) is shown in Fig. 8. This figure shows which pairs of $(\dot{\gamma}/\dot{\gamma}_a)$ — and (t_s/τ) — values are needed for chosen (parameter-)values of Φ. The horizontal sections in Fig. 8 indicate equilibrium between growth and decay. In general, the value of τ can be determined fairly accurately, independent of the (uncertain) choice for Φ_{crit}. In fact, values for Φ_{crit} very close to unity, where the bend in the theoretical curves shifts to values of t_s/τ much smaller than unity, do not seem probable, the more so as such a choice of Φ_{crit} becomes very critical: Tremendous changes in t_s/τ would be caused by tiny variations in Φ_{crit}. Interestingly, in Fig. 7 the bend occurs at shearing times of the order of 7 s. Independent from the birefringence decay of the quenched surface layers, Wippel evaluted a value of τ = 7.27 s. This is in nice agreement for the shearing temperature of 200°C of PP Daplen KS 10.

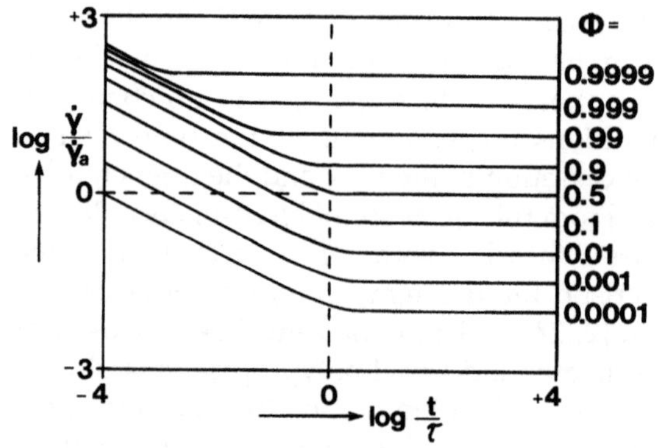

Fig. 8. Parameter presentation of the isothermal solution of Eq. (15) for several values of Φ_{crit}.

After reading this theoretical section, one may wonder how the function Φ fits into the picture of the rate equations. The answer is that Φ can take over the function of φ_3 or of $\dot{\varphi}_3$ in Eqs. (5), dependent on the crystallization mechanism considered. This will be explained in more detail at the end of the next section.

Short isothermal shearing

In the just described quenching experiments flow was practically unperturbed, because solidifcation occurred only after the quench, which followed the cessation of flow. In the experiments described in the following, flow was again assumed to be undisturbed by incipient crystallization. But this goal was reached in another way. Short-term isothermal shearing and subsequent (rather long-term) development of induced crystallization was investigated at low levels of supercooling, where normal spherulitic crystallization occurs only after some hours.

Two methods of observations were used [31, 2]:

a) Investigation of cross-sections, as before;
b) Optical monitoring of the development of the crystallization as induced by short-term shearing.

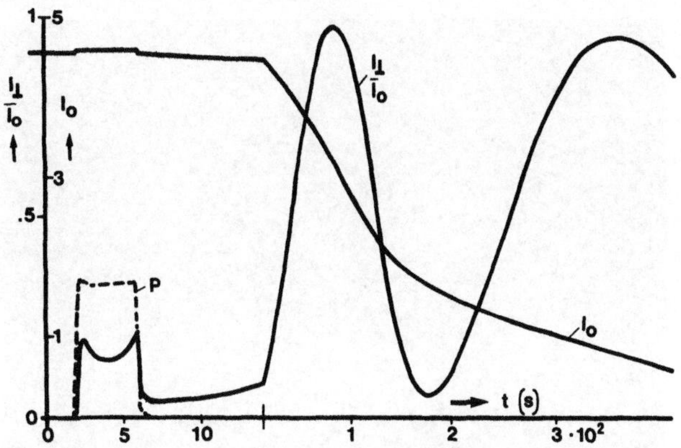

Fig. 9. Liedauer's method for monitoring the isothermal development of shear-induced crystallization after short-term shearing, according to [2] (courtesy of Pergamon Press)

This point is explained in Fig. 9. This figure shows the method developed by Liedauer. The lengthy rectangular duct was first filled with polymer melt by extrusion at a temperature well above the melting point. After flow was stopped and an appropriate time for relaxation had elaspsed, the duct was cooled to the chosen temperature level below the melting point. When the experiment was started a "pressure box" was applied to the duct entrance, which means that a constant extrusion pressure was applied for a sharply determined time span. After this time the increase of the turbidity and of the optical path difference was observed through glass windows mounted into the large walls of the duct near its exit. Two quantities were measured continuously: The intensity I_0 and the ratio of the intensity $I_\perp$ as obtained with crossed polars, divided by the total intensity I_0, $(= I_\perp + I_\parallel)$, viz.:

$$I_\perp / I_0 = \sin^2(\pi \Gamma / \lambda) \ . \tag{16}$$

Here, Γ is the optical retardation and λ the wave length. The analyser was under 45° with the direction, of the duct; the polarizer was rotated with a frequency of 30 Hz around an axis in its plane. (Details will be published in the near future [31].) Turbidity and path difference are of interest, as will become obvious from the next figure. With respect to Fig. 9, one observes that the "pressure box" is indicated by a direct pressure measurement and by a "box" in the retardation which, however, is a little deformed, indicating an overshoot in the beginning and some build-up of anisotropic structures at the end. But it mainly seems to be a consequence of flow birefringence which disappears when the shear stress becomes zero. After a little more than 10 s the speed of the recorder paper is reduced. This causes the kinks in the curves. One notices that, after a shearing time of roughly 3.5 s, the first maximum in $I_\perp / I_0$ is reached only after about 100 s.

This maximum indicates that the retardation has reached half the wave length. At the same time the intensity I_0 has fallen to about half of its orignal value.

In Fig. 10 fine grained surface layers can be observed for a sample of PP Daplen KS 10, as obtained after a shearing for $t_s = 10$ s at $\dot{\gamma}_w = 43\ \mathrm{s}^{-1}$ and $T = 150$°C. ($\dot{\gamma}_w$ indicates the maximum shear rate at the large duct walls.) Here, we notice that after moderate shearing only layers of a fine grained structure but no birefringent layers are obtained. This effect was overlooked in the previous shearing experiment because of the fact that, with a rapid quench, the whole sample shows a fine-grained structure. Here, the final quench occurred only after about 700 s and could be carried out slowly. With increased intensity of shearing ($\dot{\gamma}_w = 68\ \mathrm{s}^{-1}$ and $t_s = 15{,}6$ s at 150°C) two types of layers are obtained, as shown in Fig. 11. The birefringent structure, however, appears to be superimposed on the fine-grained structure. If the birefringence is ex-

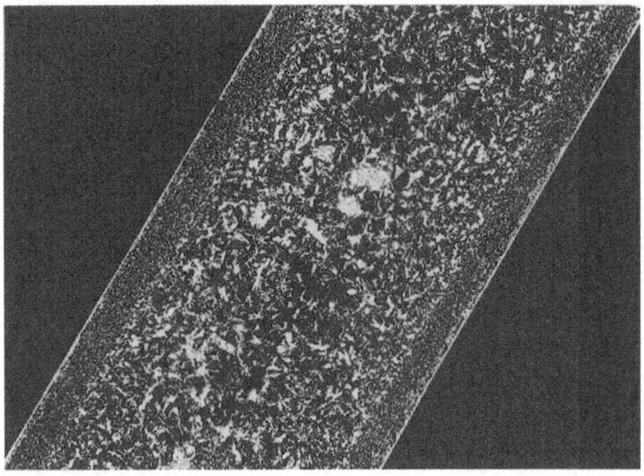

Fig. 10. Fine-grained surface layers in a sample of polypropylene Daplen KS 10 after short-term shearing of t_s = 10 s at a shear rate of $\dot{\gamma}_w$ = 43 s^{-1} at the large duct walls and at a temperature of T = 150°C

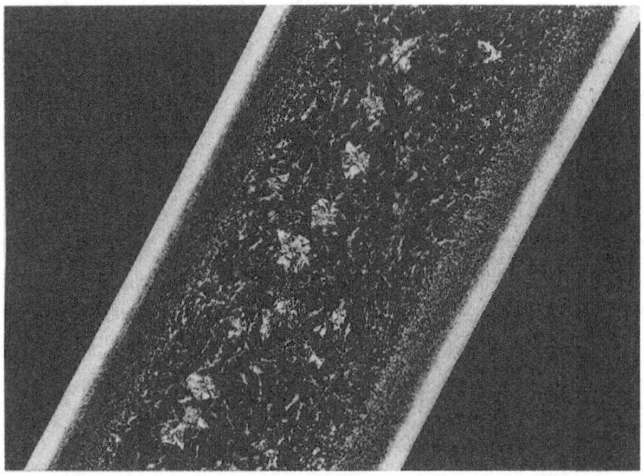

Fig. 11. Two types of surface layers in a sample of polypropylene Daplen KS 10 after a more intensive shearing (t_s = 15, 6 s, $\dot{\gamma}_w$ = 68 s^{-1}, T = 150°C)

tinguished by rotation of the crossed polars (or the microscope table), a fine-grained structure becomes observable also close to the surface. The observed increase of turbidity may mainly be due to the development of the fine-grained structure. The layer thicknesses appear to be independent of the time of the prolonged development of the structures and, thus, also of the growth of the birefringence. The latter points to the development of highly oriented crystals, their extinction positions being parallel and perpendicular to the sample surface. Therefore, the moment of the final quench has no influence on the thickness of the layers. This means that the precursors have all been formed during the short periods of shearing.

In Fig. 12 a double logarithmic plot is given of the shear rates at which shear induced crystallization occurred at shearing times t_s. Measurements were carried out at 157°, 150°, and 143°C with PP Daplen KS 10. There are two types of curves: dashed lines and full lines. Below these curves no shear induced crystallization can be observed. Dashed lines indicate the occurrence of fine-grained structures, full lines that of highly oriented lamellar textures. The spread of the actual measuring points is similar to the one shown in Fig. 7. Also, the calculation of the shear rates at various distances from the central plane occurred in a similar way. In the present case, however, the layers show constant thickness along the duct, as no entrance flow is considered. This eases the evaluation. Nevertheless, experiments with varying through-puts were of great use, since with a lower through-put a certain rate of shear occurs more closely to the wall and vice versa.

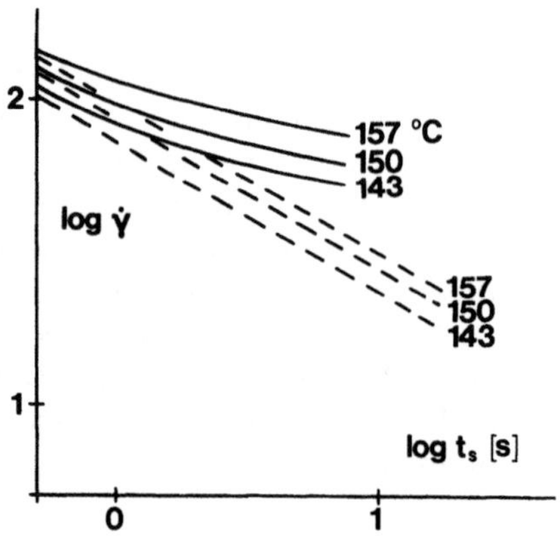

Fig. 12. Double-logarithmic plot of critical shear rate against shearing time for three shearing temperatures: 157°, 150°, and 143°C for polypropylene Daplen KS 10. Solid lines indicate oriented lamellar crystallization, dotted lines indicate fine-grained structure

According to Fig. 12, there seem no great variation in the shearing times where the lines for the oriented layers start to deviate from the lines with slope —1/2. The lines are assumed to have a slope

—1/2 at still lower shearing times. Under this condition any deviations of the sort shown in the figure seem to indicate an incipient bend towards a horizontal line. Thus, the respective relaxation times seem to be not much larger than the one found for the shearing at 200 °C. This seems surprising in view of the results shown in Fig. 6, according to which — by extrapolation down to 150 °C, the relaxation times should become enormous. However, the accuracy of the measurements may be insufficient to indicate the said bend. Conversely, it can be that in a melt, after cessation of flow, the relaxation times are much higher than in a flow situation. In fact, the behavior of the material is certainly highly non-linear which justifies such an extra assumption.

One thing is clear: the dashed lines show a slope of —1/2 over the whole range of shear rates and shearing times. The pertinent relaxation times must be very high. This seems to be in accordance with the observation that filling at a much higher temperature is critical. Even if no subsequent shearing occurs at a lower temperature, some fine-grained texture near the wall is very persistent and relaxation before cooling seems insufficient.

Fine-grained crystallization was observed previously when shearing was realized in a rotational viscometer. For isotactic poly(butene-1), Wolkowicz [32] reported a multiplication of the number of nuclei. Sherwood et al. [33] reported Avrami exponents increasing with shear rate up to about 7 for poly(ethylene oxide). Identifying our Φ with φ_3, one can, in principle, explain this phenomenon with our theory [2, 29].

Finally, the development of the optical retardation Γ after short-term shearing should be scrutinized. In Fig. 13 results are depicted for a shear rate at the large duct walls of $\dot{\gamma}_w = 169\ \mathrm{s}^{-1}$ at 150 °C (for PP Daplen KS 10) [31]. The shearing times are varied by a factor of 5. All these times, however, are extremely short compared with the times over which the subsequent development was monitored. If the birefringence is assumed to be proportional to the lateral overgrowth on lamellar precursors, such a one-dimensional growth is manifested by the experiments: All the curves of Fig. 13 start with a linear section, indicating an Avrami exponent of 1. The initial slopes appear to be proportional to

$$\Phi^2 = \left[\left(\frac{\dot{\gamma}}{\dot{\gamma}_a}\right)^2 \frac{t_s}{\tau}\right]^2 . \tag{17}$$

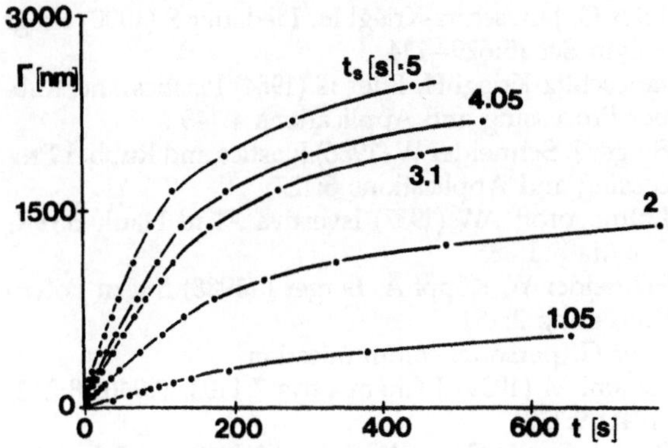

Fig. 13. Development of the optical retardation for polypropylene Daplen KS 10 at 150 °C after shearing at $\dot{\gamma}_w = 169\ \mathrm{s}^{-1}$ for the shearing times indicated in the graph. Note the extended time of observation

This can be seen from a double logarithmic plot [34], where Φ (or Φ^2) apparently takes over the role of φ_3 in Eq. (5a). A detailed presentation and discussion will be given elsewhere.

The present author apologizes for the sporadic citation of literature on shear-induced crystallization. The reader may find the review in [2] beneficial. In the present context, the author had to restrict himself to the presentation of a general context which became recognizable only recently.

Acknowledgement

The author acknowledges financial support by the Austrian "Fonds zur Förderung der wissenschaftlichen Forschung," obtained during a 5-year national working party (S 33) on the "Influence of Molecular Structure and Processing Parameters on the Properties of Moulded Plastic Parts" (project no. S 33 02) and in subsequent projects P7423 and P7424 CHE. He would also like to mention the indirect support provided through a more practical project 5/570, as sponsored by the Austrian "Forschungsförderungsfonds für die Gewerbliche Wirtschaft" together with Petrochemie Danubia. He thanks his collaborators, Drs. G. Eder and C. H. Wu, and his doctoral students E. Ratajski, S. Liedauer, and H. Wippel and M. Lipp for constructing the instruments.

References

1. Final Report, National Working Party S 33 (1991) "Influence of Molecular Structure and Processing Parameters on the Properties of Moulded Plastic Parts", H. Janeschitz-Kriegl (ed) Inst of Chemistry, Johannes Kepler University Linz, A-4040 Linz

2. Eder G, Janeschitz-Kriegl H, Liedauer S (1990) Progr Polym Sci 15:629–714
3. Janeschitz-Kriegl H, Eder G (1984) Plastics and Rubber Processing and Applications 4:145
4. Berger J, Schneider W (1986) Plastics and Rubber Processing and Applications 6:127
5. Kolmogoroff AW (1937) Isvestiya Akad Nauk SSSR, Ser Math 1:355
6. Schneider W, Köppl A, Berger J (1988) Intern Polym Processing 2:151
7. Eder G, personal communication
8. Avrami M (1939) J Chem Phys 7:1103; (1940) 8:212; (1941) 9:177
9. Boon J, Challa G, van Krevelen DW (1968) J Polym Sci A2 6:1791
10. Boon J, Challa G, van Krevelen DW (1968) J Polym Sci A2 6:1835
11. Mandelkern L (1964) "Crystallization in Polymers", McGraw Hill, New York
12. Wunderlich B (1973) "Macromolecular Physics", Acad Press New York, London, Vol 1, pp 214, 464
13. Basset DC (1981) "Principles of Polymer Morphology", Cambridge Univ Press, p 218
14. Nakamura K, Watanabe T, Katayama K, Amano T (1972) J Appl Polym Sci 16:1077
15. Janeschitz-Kriegl H, Eder G (1991) "Progress in the Understanding of Crystallization Kinetics and Polymer Processing Conditions", 7th Annual Meeting, Polymer Processing Society, April 1991, Hamilton, Ontario, Canada
16. van Krevelen DW (1978) Chimia 32:279
17. Magill HJ, Li HM, Gandica A (1973) J Crystal Growth 19:361
18. Magill HJ (1964) J Appl Phys 35:3249; (1967) J Polym Sci A2 5:89
19. Magill HJ (1968) Polym Letters 6:853
20. Janeschitz-Kriegl H, Eder G, Krobath G, Liedauer S (1987) J Non-Newtonian Fluid Mech 23:107
21. Eder G, Paulik Ch, Wu CH, "Polymer Crystallization Dynamics, as Reflected by DSC", paper III, Colloid Polym Sci (in preparation)
22. Billon N, Haudin J-M (1991) "Determination of Nucleation Rate in Polymers using Isothermal Crystallization and Computer Simulation", Poster at 23rd Europhysics Conference on Macromolecular Physics, Stockholm, May 1991
23. Janeschitz-Kriegl H, Wippel H, Paulik Ch, "Polymer Crystallization Dynamics, as Reflected by DSC", paper I, Colloid Polym Sci (in preparation)
24. Wu CH, Janeschitz-Kriegl H, ibid paper II, Colloid Polym Sci (in preparation)
25. Janeschitz-Kriegl H, Wimberger-Friedl R, Krobath G, Liedauer S (1987) Kautschuk + Gummi, Kunststoffe 40:301
26. Kantz MR, Newman HD, Stigale FH (1972) J Appl Polym Sci 16:1249
27. Wippel H (1989) report delivered in fulfillment of the requirments for the degree of Diplom-Ingenieur, Linz Univ
28. Bird RB, Armstrong RC, Hassager O (1987) "Dynamics of Polymeric Liquids", Vol 1, 2nd ed, John Wiley, NY, p 172
29. Eder G, Janeschitz-Kriegl H (1988) Colloid Polym Sci 266:1087
30. Eder G, Janeschitz-Kriegl H, Krobath G (1989) Progr Colloid Polym Sci 80:1
31. Liedauer S, Doctoral Thesis Linz Univ (in preparation)
32. Wolkowicz MD (1978) J Polym Sci Polym Symp 63:365
33. Sherwood CH, Price FP, Stein RS (1978) J Polym Sci Polym Symp 63:77
34. Eder G, Janeschitz-Kriegl H, Liedauer S (1991) "Influence of Flow on the Crstallization Kinetics of Polymers", this volume

Author's address:

H. Janeschitz-Kriegl
Inst. of Chemistry
Linz University
A-4040 Linz, Austria

Progress in Colloid & Polymer Science Progr Colloid Polym Sci 87:129—131 (1992)

Influence of flow on the crystallization kinetics of polymers

G. Eder, H. Janeschitz-Kriegl, and S. Liedauer

Institute of Chemistry, Linz University, Austria

Abstract: In the cross-sections of injection-molded parts made from crystallizable polymers there are, in general, at least two types of morphologies. In the interior, spherulitic crystallization predominates, as also occurs with solidification of quiescent melts. Near the surfaces we find highly oriented, often optically structureless layers caused by shear-induced crystallization. — With the aid of an isothermal extrusion experiment at just a few degrees below the thermodynamic melting temperature, where the typical crystallization time for quiescent crystallization has the magnitude of some hours, it is possible to trace the shear-induced crystallization process. Polypropylene is extruded during a short shearing time through a rectangular duct, and the birefringence across the end of the duct is measured during and after flow. If the shearing time is not too large, nucleation and growth process of the shear-induced layers are separated. Nucleation of thin precursor sheets takes place during shear, whereas after stop of flow one gets a process of Avrami index 1 due to one-dimensional gowth upon the sheets. By variation of the experimental parameters it is possible to trace the nucleation process during shear.

Key words: Flow-induced crystallization; shear flow; nucleation; crystallization kinetics; polypropylene

Introduction

The cross-section of injection-molded parts made from crystallizable polymers show at least two types of morphologies. In the interior spherulitic crystallized structures are found as also after solidification of quiescent melts. Near the surfaces are highly oriented, optically structureless layers caused by shear-induced crystallization.

It is well known that small rates of deformation may dramatically enlarge the number of crystallites [1], leading to a spherulitic morphology as in the case of crystallization under conditions of rest. Sometimes, with stronger deformation conditions, also shish kebab structures have been found with sheared polymer melts [2]. These nucleating rows give rise to a two-dimensional lateral growth of cylindrical crystallites. If shear rate and total shear are further increased, we have a crystallization process which may be viewed as one-dimensional growth on nucleating areas in the polymer.

Crystallization kinetic methods

Every crystallization process should be describable by means of a nucleation and a growth process. The problem of covering the three-dimensional space by growing spheres nucleating with a rate $a(t)$ per unit of volume and growing with a speed $G(t)$ was solved first by Kolmogoroff [3]. He arrived at

$$\xi(t) = 1 - e^{-\frac{4\pi}{3}\int_{-\infty}^{t} da(s)\left[\int_{s}^{t} du\, G(u)\right]^3}, \tag{1}$$

where ξ is the degree of covering of space by the growing spherulites.

The same procedure can be used to derive equations for the problem of covering the space with developing sheets in the material growing one-dimensionally. In this case, we have

$$\xi(t) = 1 - e^{-\int_{-\infty}^{t} dA(s)\left[\int_{s}^{t} du\, G(u)\right]^1}, \tag{2}$$

where $A(t)$ is the interior nucleation area per unit of volume. Such an area must be thought to be the surface area of extremely thin sheets of precursors. The inverse of this area is then half the mean distance between these sheets. The growth speed of the crystallites $G(t)$ should depend on time t only indirectly via the time-dependent temperature.

The generation of the nucleating areas may be described by a differential equation with a creation term due to the flow and a decay term due to Brownian motion [4, 5], the latter being important only at temperatures well above the melting temperature.

Model experiment — isothermal extrusion below T_m

Polyproylene melt is extruded isothermally below the melting temperature T_m through a rectangular duct during a short shearing time t_s. The birefringence across the end of the duct is measured during and after flow as a function of time t. This is an on-line measurement of the shear-induced crystallization process alone, since the temperatures are chosen so that the typical crystallization time for quiescent crystallization has the magnitude of hours.

This model experiment enables us to separate nucleation and growth process. For small shearing times t_s the nucleation areas are formed within this period, whereas the growth process takes place on a much longer time scale. In our case the above equation simplifies to

$$\xi(t,x) = 1 - e^{-G(T)A(t_s;\dot{\gamma}(x))t} , \tag{3}$$

where $\dot{\gamma}(x)$ is the local shear rate. For small times t, which have nonetheless to be large as compared with the shearing time t_s, the degree of covering $\xi(t)$ is linear in t, which means that the Avrami index of this process is equal to 1. The quantity $A(t_s;\dot{\gamma}(x))$ can be traced through variation of the conditions of our experiment, viz. period and speed of extrusion.

If the growing crystallites are made responsible for the measured path difference of polarized light beams at the end of the duct, we can write

$$\Gamma(t) = \int_{-H/2}^{H/2} dx\,\xi(t,x)\Delta n_c(t_s;\dot{\gamma}(x)) , \tag{4}$$

where Δn_c is the birefringence of the crystallites and H is the duct height. For small time t we have, therefore,

$$\Gamma(t) = 2tG(T)\int_{0}^{H/2} dx(A\Delta n_c)(t_s;\dot{\gamma}(x)) , \tag{5}$$

which is linear in time (compare Fig. 1). According to the theory of shear-induced crystallization [4] the quantity $(A\Delta n_c)$ should depend on the precursor-function Φ, which fulfills the differential equation

$$\tau\dot{\Phi} = (\dot{\gamma}/\dot{\gamma}_a)^2(1 - \Phi) - \Phi . \tag{6}$$

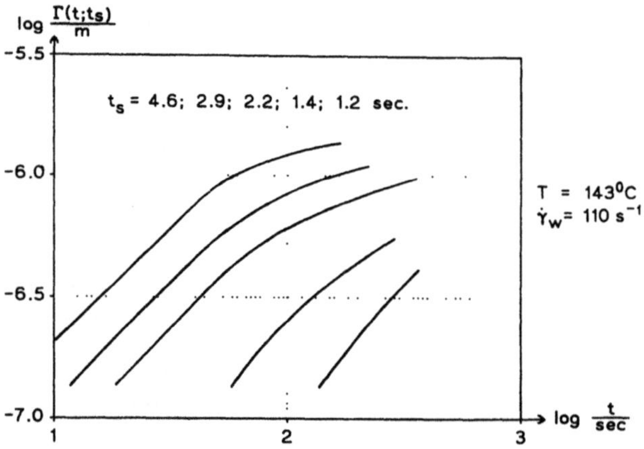

Fig. 1. Path difference of polarized light beams of wave length λ = 546 nm for a polypropylene Daplen KS10 melt at a temperature of T = 143 °C and shear rate at the wall $\dot{\gamma}_w$ = 110 s^{-1} for various shearing times t_s as function of time t

For this experiment the decay term is negligible and the solution therefore reads

$$\Phi = \frac{\dot{\gamma}^2 t_s}{\dot{\gamma}_a^2\tau} . \tag{7}$$

Assuming as simple a relationship between $(A\Delta n_c)$ and Φ as possible,

$$A\Delta n_c = K\Phi^m , \tag{8}$$

with constants K and m, we are able to calculate the time needed for Γ to reach the value $\lambda/2$. This value is chosen because it lies, for all experimental results, in the linear region of $\Gamma(t)$.

Using a power law model for the rheological equation of state (i.e., $\dot{\gamma}(x) = \dot{\gamma}_w \left(\frac{2x}{H}\right)^{1/n}$ with wall shear rate $\dot{\gamma}_w$ and power law index n), we get

$$t_{\lambda/2} = \varkappa(\dot{\gamma}_w^2 t_s)^{-m} , \tag{9}$$

with a constant $\varkappa$ depending on λ, H, K, m, n and the crystallization kinetic constants $\dot{\gamma}_a$ and τ only. Figure 2 shows that this relation is well fulfilled for $m = 2$. The parallel lines in this figure have slope -2 and a mutual distance of $4\,\Delta(\log(\dot{\gamma}_w))$.

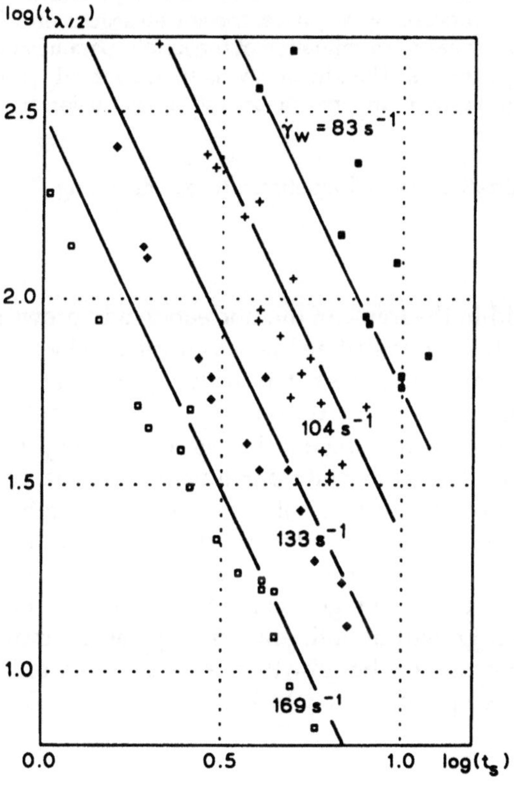

Fig. 2. Time needed for the path difference Γ to reach $\lambda/2$ as function of shearing time t_s for various wall shear rates $\dot{\gamma}_w$ at a temperature of $T = 150\,^\circ\mathrm{C}$ for polypropylene Daplen KS10

In addition, the thickness D of the shear-induced layers can be calculated from the place where, at the time of the final quench (after 12 min), the crystallinity had some constant value. Finally, we have

$$D = \frac{H}{2}\left(1 - (C\dot{\gamma}_w^2 t_s)^{-n/2}\right) , \tag{10}$$

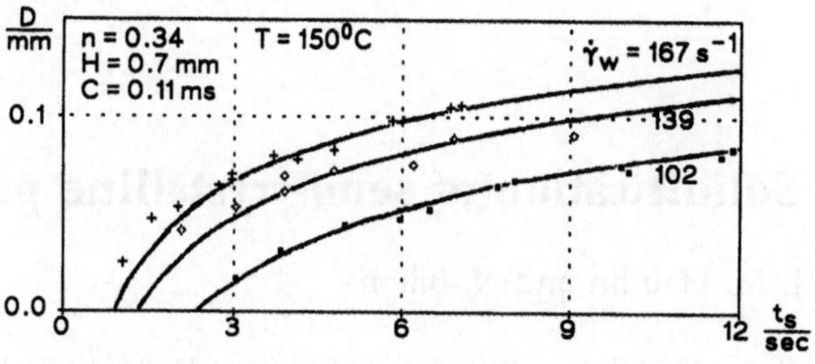

Fig. 3. Thickness of the shear-induced layer as a function of shearing time t_s and shear rate at the wall $\dot{\gamma}_w$ in a rectangular duct. Solid lines calculated according to Eq. (10)

with the temperature-dependent time C and power law index n (see Fig. 3).

Conclusion

With the aid of a very simple geometrical model, viz. formation of thin precursor sheets growing subsequently in one dimension only, the build-up of shear-induced layers after isothermal short-term shearing just below the melting temperature could be successfully described. This experiment makes it possible to separate nucleation (during shear) and growth process (after shear). Through on-line birefringence measurements the Avrami index of this crystallization process is found to be 1, due to one-dimensional growth of the crystalline regions. According to a simple theory published before [4], the quantity $\dot{\gamma}^2 t_s$ determines the speed of crystallization after shear at shear rate $\dot{\gamma}$ and shearing time t_s.

References

1. Wolkowicz MD (1978) J Polym Sci, Polym Symp 63:35
2. Ulrich RD, Price FP (1976) J Appl Polym Sci 20:1077
3. Kolmogoroff AN (1937) Isvest Akad Nauk, Ser Math 1:355
4. Eder G, Janeschitz-Kriegl H (1988) Colloid Polym Sci 266:1987
5. Eder G, Janeschitz-Kriegl H, Liedauer S (1990) Prog Polym Sci 15:629

Authors' address:

Dr. Gerhard Eder
Institute of Chemistry
Linz University
A-4040 Linz, Austria

Solidification of semi-crystalline polymers during melt processing*)

J. M. Haudin and N. Billon

Centre de Mise en Forme des Matériaux, Unité de Recherche Associée au CNRS n° 1374,
Ecole Nationale Supérieure des Mines de Paris, Sophia Antipolis, Valbonne, France

Abstract: The classical theories for overall crystallization kinetics are first reviewed, then discussed and extended. To model the crystallization phenomena during polymer processing, we have chosen an extension of Ozawa's theory which allows an easy determination of the kinetic parameters from anisothermal DSC experiments. The kinetic law is then coupled with the energy equation. Two applications are presented: cast-film extrusion and injection molding.

Key words: Polymer; crystallization; overall kinetics; process modeling

Introduction

A powerful tool for the understanding of polymer processing is modeling. Generally, the first models were purely thermomechanical ones. Recently, more and more models for the processing of semi-crystalline polymers have been proposed to take into account the crystallization phenomena in a more or less sophisticated manner.

The basic idea is to introduce a crystallization law (overall crystallization kinetics) into a thermomechanical model of the process. Several approaches can be used to describe polymer solidification by crystallization. In fact, most of them are based on the Kolmogoroff-Avrami-Evans equation. General and approximate forms of this equation will be presented first. Then the coupling between the kinetic law and the energy equation will be explained. Finally, two applications will be given: cast-film extrusion and injection molding.

Overall crystallization kinetics

General hypotheses

The models for overall crystallization kinetics are generally based on the same hypotheses. They are built within the frame of the nucleation and growth approach. The crystalline phase is nucleated by tiny nuclei, which already exist in the liquid, with an initial number per unit volume N_0. These potential nuclei become growing entities, as soon as they are activated. The probability per unit time of the occurrence of an activation, also named activation frequency q, depends only on the crystallization temperature. An entity grows in all the available directions till it impinges on another entity. At that moment, growth of both entities stops at the impingement point. The growth rate G depends only on the temperature at which crystallization takes place.

The classical form of the Kolmogoroff-Avrami-Evans theory

The original papers of Avrami [1—3] and Evans [4] are not always clear and are sometimes restrictive. For instance, Evans' initial theory is isothermal and no general demonstration is given. In fact, it can be demonstrated in a rigorous way and generalized [5—7]. Furthermore, it can be shown that Avrami's and Evans' theories are formally equivalent in the case where the potential nuclei are randomly distributed [5—7]. Eventually, it must be noted that several years before, Kolmogoroff [8] had obtained

*) Invited Lecture at the 23rd Europhysics Conference on Macromolecular Physics, Stockholm, May 1991.

the same results as Avrami and Evans, using a statistical approach. Therefore, these initial theories should be described by the generic term "Kolmogoroff-Avrami-Evans" (KAE) theory. All of them assume that the crystallizing volume is constant and infinite.

The KAE theory gives the following expression for the volume volume fraction $\alpha(t)$ transformed at time t:

$$\alpha(t) = 1 - \exp[-E(t)] \,. \tag{1}$$

In the case of three-dimensional crystallization, α (which ranges from 0 to 1) is the ratio of the volume overlapped by crystalline or semi-crystalline entities to the total volume of the specimen. The calculation is valid for special shapes of entities (spheres, disks in parallel planes, parallel rods). $E(t)$ is the total number of entities, which may have passed over a given point in the medium at time t, if these entities are assumed to nucleate and grow independently of each other. $E(t)$ is equal to the "extended" transformed volume fraction introduced by Avrami [2].

Generalization of Evans' approach

Evans' approach can be easily generalized to the case where the initial number of potential nuclei per unit volume is defined *locally* by a function $N_0(\rho,\theta,\phi)$ of the spherical coordinates ρ,θ,ϕ [5–7]. If the temperature does not depend on ρ, θ, and ϕ, the activation frequency q and the growth rate G depend only on time. Then,

$$E(t) = \int_0^{R(t)} \int_{\theta=0}^{\pi} \int_{\phi=0}^{2\pi} N_0(\rho,\theta,\phi)[1 - \exp[-\eta(\tau)]]\rho^2 \cdot \sin\theta d\phi d\theta d\rho \,, \tag{2}$$

where

$$R(t) = \int_0^t G(u)du \tag{3}$$

$$\rho(t,\tau) = \int_\tau^t G(u)du \tag{4}$$

$$\eta(\tau) = \int_0^\tau q(u)du \,. \tag{5}$$

Equation (2) is a general equation which makes it possible to determine the crystallization kinetics whatever the cooling conditions, under the following assumptions:

- isovolumic assumption (as in the initial KAE theory), but now the volume is not necessarily infinite [6];
- special shapes of the entities: spheres, disks or rods (as in the initial theory);
- locally homogeneous temperature, i.e., homogeneous at the scale of crystallization phenomena.

Nevertheless, the use of this equation requires knowledge of N_0, q, and G. It is generally difficult to determine the nucleation parameters N_0 and q [9]. Therefore, additional assumptions are frequently made and approximate expressions are currently used.

Additional assumptions

An important additional assumption is the isokinetic assumption, first proposed by Avrami [1]: G/q is assumed to be constant. It follows that

$$\rho(t,\tau) = \frac{G}{q}[\eta(t) - \eta(\tau)] \,, \tag{6}$$

and

$$E(t) = \left(\frac{G}{q}\right)^3 \int_0^{\eta(t)} \int_{\theta=0}^{\pi} \int_{\phi=0}^{2\pi} N_0(\rho,\theta,\phi) \cdot [1 - \exp[-\eta(\tau)]] \cdot [\eta(t) - \eta(\tau)]^2 \sin\theta d\phi\, d\theta d\eta(\tau) \tag{7}$$

In the case of a uniform intitial distribution of potential nuclei, Eq. (7) reduces to

$$E(t) = 8\pi N_0 \left(\frac{G}{q}\right)^3 f(\eta(t)) \,, \tag{8}$$

where

$$f(x) = \exp(-x) - 1 + x - \frac{x^2}{2} + \frac{x^3}{6} \,. \tag{9}$$

Furthermore, if limiting cases of nucleation (i.e., initial (q large) or sporadic in time (N_0 large, q small and constant nucleation rate) are considered, Eq. (8) becomes

$$E(t) = k\eta^n(t) \,, \tag{10}$$

where the Avrami exponent n has an integer value and k is a constant.

Approximate expressions

In view of the lack of experimental data concerning N_0 and q, many authors have been obliged to use in practice simplified forms of the KAE theory, even if they had initially established more general equations. The most popular simplified expression of the KAE equation is

$$\alpha(t) = 1 - \exp[-k\eta^n(t)] \ . \tag{11}$$

It was shown above that the Avrami exponent n is an integer in special cases of nucleation. However, it is often considered that Eq. (11) remains a good approximation, when nucleation is neither purely instantaneous, nor purely sporadic. In such cases, n has non-integer values.

Different variants of Eq. (11) are proposed in the literature, depending on the procedure used to determine the kinetic parameters: isothermal or non-isothermal. A first route (proposed, for instance, by Nakamura et al. [10—12]) is to use isothermal data to model anisothermal crystallizations. In the case of an isothermal crystallization

$$\alpha(t) = 1 - \exp[-k_i t^n] \ , \tag{12}$$

where k_i is a parameter which depends only on the crystallization temperature. If

$$K(u) = k^{1/n} q(u) = k_i^{1/n}(T(u)) \ , \tag{13}$$

Eq. (11) can be written

$$\alpha(t) = 1 - \exp\left[-\left(\int_0^t K(u)du\right)^n\right] = 1 - \exp\left[-\ln 2\left(\int_0^t \frac{du}{t_{0.5}(T(u))}\right)^n\right] \ , \tag{14}$$

where $t_{0.5}$ is the half-crystallization time under isothermal conditions. It can be measured as a function of the crystallization temperature or given by a theoretical expression: Arrhenius-type [13], Gaussian-type [14, 15] or WLF-type [14, 16].

Another way is to use crystallizations performed at a constant cooling rate $\dot{T}$ ($\dot{T} < 0$ for a cooling). Then the transformed volume fraction is given by

$$\alpha(t) = 1 - \exp\left[-\frac{k}{|\dot{T}|^n}\left(\int_{T(t)}^{T_m^0} q(\Gamma)d\Gamma\right)^n\right] \ , \tag{15}$$

where T_m^0 is the equilibrium melting temperature. This expression is similar to that proposed by Ozawa [17] in the two limiting cases of nucleation (instantaneous and sporadic in time):

$$\alpha(t) = 1 - \exp\left[-\frac{\chi(T)}{|\dot{T}|^n}\right] \tag{16}$$

so that

$$\frac{\partial \chi^{1/n}(T)}{\partial T} = -k^{1/n} q(T) = -k_i^{1/n}(T) \ . \tag{17}$$

Choice of a kinetic law

Although we have already developed a more complete formulation [5—7], we have used for the modeling of polymer processing a simplified one, based on Ozawa's equation (16). The major advantage of this simplified form is that it requires the experimental determination of only two parameters: the Avrami exponent n and the function $\chi(T)$. This determination is classically made by differential scanning calorimetry (DSC) during crystallizations at a constant cooling rate, which is a major difference with respect to previous approaches based on isothermal measurements.

The difference does not result from theoretical considerations since Eq. (17) gives the correspondence between χ and k_i. This equation has been experimentally checked in the case of medium density polyethylene [18]. The main reason for our choice is experimental: it is likely that, during polymer processing, crystallization will occur at low temperatures, since the cooling is generally rapid. Isothermal crystallizations at low temperatures are so rapid that measurements are often impossible. On the contrary, anisothermal measurements do not have such limitations and allow to characterize polymers in the range of temperature really encountered by the material.

Billon [5, 19] has extended Ozawa's equation to the case where the cooling rate is time-dependent. In such a case the cooling is decomposed in linear steps: the cooling rate a_j is assumed to be constant in the $[T_{j-1}, T_j]$ temperature interval. The model is also valid in the case of partial heating ($a_j > 0$) and in the case of an isothermal step, e.g., between t_i and t_k. As a consequence, the general form of E, taking into account all the possible cases (cooling, heating, isothermal step), can be written:

$$[E(t_p)]^{1/n} = \left[\sum_{j=1}^{i}\left|\frac{\chi^{1/n}(T_j) - \chi^{1/n}(T_{j-1})}{a_j}\right|\right] - \left[\frac{\partial\chi^{1/n}(T_i)}{\partial T}(t_k - t_i)\right] + \left[\sum_{j=k+1}^{p}\left|\frac{\chi^{1/n}(T_j) - \chi^{1/n}(T_{j-1})}{a_j}\right|\right] . \quad (18)$$

This general equation has been used to model polymer processing. Dufossé has shown that our procedure gives the same results as the previous ones, in which the cooling sequence is decomposed into isothermal steps [18, 20].

Coupling with the energy equation

In our approach [5, 19], the energy equation is written in terms of the enthalpy per unit mass H:

$$\frac{d(\rho H)}{dt} = \frac{\partial(\rho H)}{\partial t} + \vec{V}\cdot \vec{\text{grad}}(\rho H) = \text{div}(k\,\vec{\text{grad}}\,T) + \dot{W} , \quad (19)$$

where $\vec{V}$ and $\dot{W}$ are the polymer velocity and the power dissipated per unit volume by viscous forces, respectively. By consistency with the crystallization model (isovolumic assumption), the polymer density ρ is assumed to be constant during the cooling; k is the polymer thermal conductivity.

H depends on the temperature T and the crystalline fraction x_c. Let c and ΔH_0 be, respectively, the polymer specific heat and the specific enthalpy of crystallization:

$$\begin{cases} c = \dfrac{\partial H}{\partial T} \\ -\Delta H_0 = \dfrac{\partial H}{\partial x_c} \end{cases} . \quad (20)$$

Equation (19) expresses the evolution of both the temperature and the crystallinity ratio, x_c, of the polymer. The model presented in the previous section allows the calculation of the transformed volume fraction α. If the final crystallinity ratio x_f is known and constant, the crystallinity ratio x_c, at any time t, can be expressed as

$$x_c(t) = \alpha(t)x_f . \quad (21)$$

Let ΔH be the actual heat of crystallization per unit mass, as measured by DSC: $\Delta H = x_f\Delta H_0$. Equation (19) can be rewritten as

$$\rho\left(c\frac{\partial T}{\partial t} - \Delta H\frac{\partial \alpha}{\partial t}\right) + \rho\vec{V}\cdot(c\,\vec{\text{grad}}\,T - \Delta H\,\vec{\text{grad}}\,\alpha) = \text{div}(k\,\vec{\text{grad}}\,T) + \dot{W} . \quad (22)$$

Application 1. Cast-film extrusion of polypropylene

In cast-film extrusion (Fig. 1), the polymer melt is extruded through a slit die, slightly stretched for a short distance in air, and cooled on the metallic surface of a thermoregulated chill-roll. It has been shown [21–24] that there is only a small decrease of the film temperature in air (3 to 8 °C). As a consequence, the film cannot crystallize in air, but only on the roll under very unsymmetrical cooling conditions.

The crystallization of the polypropylene film on the roll has been modeled according to the procedure defined in the preceding sections [23, 24]. In such a case, the energy equation reduces to

$$u_R\frac{\partial T(x,z)}{\partial x} = \frac{k}{\rho c}\frac{\partial^2 T(x,z)}{\partial z^2} + u_R\frac{\Delta H}{c}\frac{\partial \alpha(x,z)}{\partial x} , \quad (23)$$

where the x and z coordinates correspond to the flow direction and to the film thickness, respectively; u_R is the roll velocity, and the thermal conductivity k is assumed to be constant. From DSC experiments, it has been shown that the Avrami exponent is equal to 3 and $\ln(\chi(T))$ can be fitted by a polynomial expression:

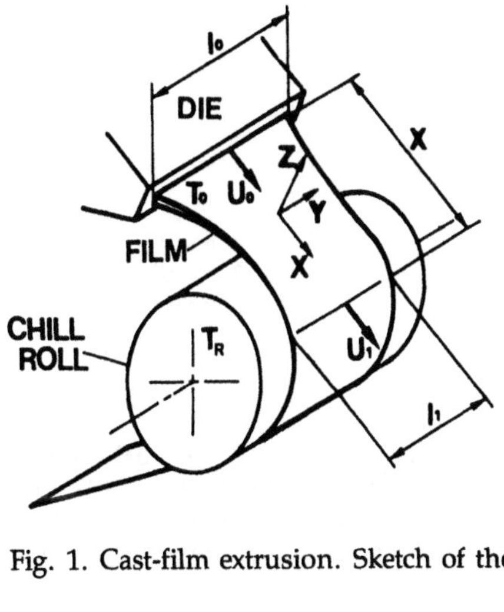

Fig. 1. Cast-film extrusion. Sketch of the process

$$\ln(\chi(T)) = 13.41 + 3.49 \cdot 10^{-2}\,T - 1.72 \cdot 10^{-3}\,T^2 + 2.09 \cdot 10^{-5}\,T^3 - 1.82 \cdot 10^{-7}\,T^4\,, \qquad (24)$$

with χ in $(K \cdot s^{-1})^n$ and T in °C.

Figure 2 illustrates the thermal conditions of crystallization in the thickness of a given PP film. The curves labeled 1, 2, 3 correspond to constant values of the transformed volume fraction: $\alpha = 0.01$ (beginning of the crystallization), $\alpha = 0.5$ (half-transformation), and $\alpha = 0.99$ (end of the crystallization). The lower zone (roll zone) begins to crystallize as soon as it is in contact with the roll, at lower temperatures than the median and the air zones, which begin to crystallize at the same temperature. The temperature associated with the half-transformation curve increases monotonically from the roll to the air zone. The end of the crystallization occurs at higher temperatures in the median zone than in the two other zones.

One can consider that the temperature characteristic of the crystallization of a zone, T_c, corresponds to $\alpha = 0.5$ in this zone. Figure 2 shows that this "crystallization temperature" increases from the roll to the air. The calculated crystallization temperatures of the three zones can be compared with an experimental determination deduced from the DSC melting curves of microtomed sections (Table 1 and [24]). The agreement between calculated and experimental values is good.

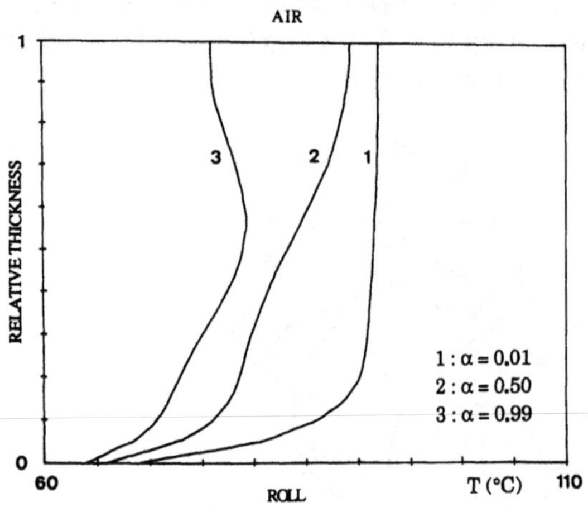

Fig. 2. Crystallization range of a 400-µm-thick polypropylene film. Extrusion temperature: 228 °C. Roll temperature: 60 °C. The curves labeled 1,2,3 correspond to fixed values of the transformed volume fraction α

Table 1. Comparison between experimentally determined and calculated crystallization temperatures

	Crystallization temperature (°C)		
Determination	Roll zone	Median zone	Air zone
DSC	80	80	90
Model	76	83	89

Application 2. Injection-molding of polyethylene

Only the filling stage will be considered here, for a disk geometry and a medium density polyethylene. The injection process is described by the classical equations of mass, momentum, and energy conservation [18]. These equations are established in a cylindrical coordinate system for an incompressible laminar flow, with appropriate initial and boundary conditions: constant flow-rate and free surface with fountain flow. The viscosity is described by a rheological law which takes into account shear rate, temperature, and pressure.

The transformed volume fraction α is computed for each material point as a function of time. It is basically described by Eq. (18), but in the case of injection molding two additional effects have been introduced:

a) *Pressure effect:* the values of n and $\chi(T)$ remain unchanged, but the crystallization temperature linearly increases with pressure (0.025 °C/bar for polyethylene).
b) *Shear effect:* the influence of the shear rate $\dot{\gamma}$ has been introduced in the following way

$$\begin{cases} \chi(T,\dot{\gamma}) = \chi(T)(1 + 3.075\,\dot{\gamma})^3 \\ n(\dot{\gamma}) = n - 1 + \dfrac{1}{1 + 0.2\,\dot{\gamma}} \end{cases} . \qquad (25)$$

These expressions were deduced by Monasse and Fryda [25] from microscopic observations of crystallizations under shear on a hot stage.

This modified kinetic law makes it possible to correctly predict the thickness of the solidified layer during the filling stage (Fig. 3). The good agreement between calculated and measured values un-

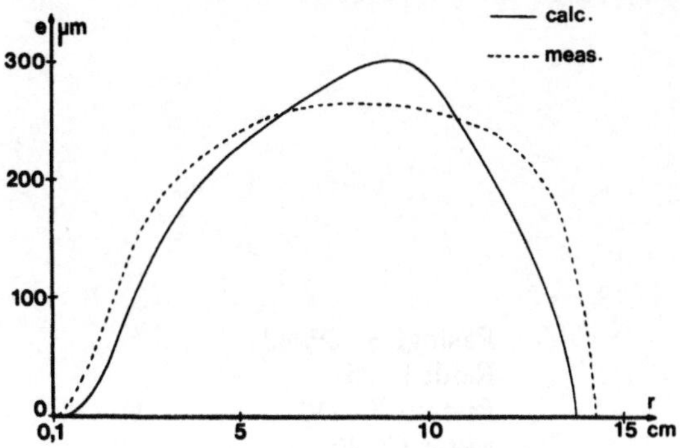

Fig. 3. Thickness of the layer solidified during the filling stage, along the radius of a polyethylene disk. Comparison between calculation and measurements. The measurements are performed on polished sections observed by optical microscopy

ambiguously shows that the shear effects must not be neglected in the modeling of such a process.

Conclusion

We have developed a general model for overall crystallization kinetics, which can be coupled with the thermomechanical models of various processes. Two examples have been presented here: cast-film extrusion and injection molding.

In its first version, this model, which is an extension of Ozawa's equation, considered only temperature effects. As shown in the present paper, it can be extended to take into account pressure or shear effects. However, the treatment of the influence of flow on crystallization remains empirical, and work is in progress in our laboratory to develop a model founded on more physical bases (extension of macromolecules, entropy variation, etc.).

Our model is derived from Ozawa's equation, i.e., from a simplified form of the kinetic law. This allows an easy determination of the relevant kinetic parameters from DSC experiments. More complete formulations are already available, but they require knowledge of the kinetic parameters N_0 and q. We are now able to propose an original method to do so [9].

References

1. Avrami M (1939) J Chem Phys 7:1103
2. Avrami M (1940) J Chem Phys 8:212
3. Avrami M (1941) J Chem Phys 9:177
4. Evans UR (1945) Trans Faraday Soc 41:365
5. Billon N (1987) Thesis, Ecole Nationale Supérieure des Mines de Paris
6. Billon N, Haudin JM (1989) Colloid Polym Sci 267:1064
7. Billon N, Haudin JM (1990) Ann Chim Fr 15:173
8. Kolmogoroff AN (1937) Izvest Akad Nauk SSR, Ser Math, 1:335
9. Billon N, Haudin JM (1991) Determination of Nucleation Rate in Polymers using Isothermal Crystallization and Computer Simulation, presented at the 23rd Europhysics Conference on Macromolecular Physics, Stockholm. Accepted for publication in Colloid Polym Sci
10. Nakamura K, Watanabe T, Katayama K, Amano T (1972) J Appl Polym Sci 16:1077
11. Nakamura K, Katayama K, Amano T (1973) J Appl Polym Sci 17:1031
12. Nakamura K, Watanabe T, Amano T, Katayama K (1974) J Appl Polym Sci 18:615
13. Markworth AJ (1983) J Appl Phys 54:3502
14. Ziabicki A (1976) Fundamentals of Fiber Formation, John Wiley and Sons, New York
15. Jeziorny A (1978) Polymer 19:1142
16. Takayunagi M, Kusumoto T (1959) Kogyo Kagaku Zasshi 62:587
17. Ozawa T (1971) Polymer 12:150
18. Dufossé C (1990) Thesis, Ecole Nationale Supérieure des Mines de Paris
19. Billon N, Barq P, Haudin JM (1991) Int Polym Processing 6:348
20. Dufossé C, Monasse B, Haudin JM (1988) Calorim Anal Therm 19:C 9.1
21. Cotto D, Duffo P, Haudin JM (1989) Int Polym Processing 4:103
22. Duffo P, Monasse B, Haudin JM (1990) Int Polym Processing 5:272
23. Duffo P (1990) Thesis, Ecole Nationale Supérieure des Mines de Paris
24. Duffo P, Monasse B, Haudin JM (1991) J Polym Eng 10:151
25. Monasse B, Fryda C (1990) Anisotropy and Crystal Growth Increase Induced by Melt Shearing of Polyethylene, presented at the 6th Annual Meeting of the Polymer Processing Society, Nice

Authors' address:

J. M. Haudin
Centre de Mise en Forme de Matériaux
Ecole Nationale Supérieure des Mines de Paris
F-06560 Valbonne, France

Author Index

Subject Index

GPSR Compliance

The European Union's (EU) General Product Safety Regulation (GPSR) is a set of rules that requires consumer products to be safe and our obligations to ensure this.

If you have any concerns about our products, you can contact us on ProductSafety@springernature.com

In case Publisher is established outside the EU, the EU authorized representative is:

Springer Nature Customer Service Center GmbH
Europaplatz 3
69115 Heidelberg, Germany

Batch number: 10578403

Printed by Printforce, the Netherlands